T0139086

Artificial Intelligence for Internet of Things

This book comprehensively discusses the essentials of the Internet of Things (IoT), machine learning algorithms, industrial and medical IoT, robotics, data analytics tools, and technologies for smart cities. It further covers fundamental concepts, advanced tools, and techniques, along with the concept of energy-efficient systems. It also highlights software and hardware interfacing into the IoT platforms and systems for better understanding. It will serve as an ideal reference text for senior undergraduates, graduate students, and academic researchers in the fields of electrical engineering, electronics and communication engineering, and computer engineering.

Features

- Covers cognitive Internet of Things and emerging network, IoT in robotics, smart cities, and health care
- Discusses major issues in the field of the IoT such as scalable and secure issues, energy-efficient, and actuator devices
- Highlights the importance of industrial and medical IoT
- Illustrates applications of the internet of things in robotics, smart grid, and smart cities
- Presents real-time examples for better understanding.

The text comprehensively discusses design principles, modernization techniques, and advanced developments in artificial intelligence. This will be helpful for senior undergraduates, graduate students, and academic researchers in diverse engineering fields including electrical, electronics and communication, and computer science.

Smart Engineering Systems: Design and Applications
Series Editor: Suman Lata Tripathi

Internet of Things
Robotic and Drone Technology
Edited by Nitin Goyal, Sharad Sharma, Arun Kumar Rana, Suman Lata Tripathi

Smart Electrical Grid System
Design Principle, Modernization, and Techniques
Edited by Krishan Arora, Suman Lata Tripathi and Sanjeevikumar Padmanaban

Artificial Intelligence, Internet of Things (IoT) and Smart Materials for Energy Applications
Edited by Mohan Lal Kolhe, Kailash J. Karande and Sampat G. Deshmukh

Artificial Intelligence for Internet of Things
Design Principle, Modernization, and Techniques
Edited by N. Thillaiarasu, Suman Lata Tripathi and V. Dhinakaran

For more information about this series, please visit: https://www.routledge.com/

Artificial Intelligence for Internet of Things

Design Principle, Modernization, and Techniques

Edited by
N. Thillaiarasu
Suman Lata Tripathi
V. Dhinakaran

CRC Press
Taylor & Francis Group
Boca Raton London New York

First edition published 2022
by CRC Press
6000 Broken Sound Parkway NW, Suite 300, Boca Raton, FL 33487-2742

and by CRC Press
4 Park Square, Milton Park, Abingdon, Oxon, OX14 4RN

CRC Press is an imprint of Taylor & Francis Group, LLC

© 2023 selection and editorial matter, N. Thillaiarasu, Suman Lata Tripathi and V. Dhinakaran individual chapters, the contributors

Reasonable efforts have been made to publish reliable data and information, but the author and publisher cannot assume responsibility for the validity of all materials or the consequences of their use. The authors and publishers have attempted to trace the copyright holders of all material reproduced in this publication and apologize to copyright holders if permission to publish in this form has not been obtained. If any copyright material has not been acknowledged please write and let us know so we may rectify in any future reprint.

Except as permitted under U.S. Copyright Law, no part of this book may be reprinted, reproduced, transmitted, or utilized in any form by any electronic, mechanical, or other means, now known or hereafter invented, including photocopying, microfilming, and recording, or in any information storage or retrieval system, without written permission from the publishers.

For permission to photocopy or use material electronically from this work, access www.copyright.com or contact the Copyright Clearance Center, Inc. (CCC), 222 Rosewood Drive, Danvers, MA 01923, 978-750-8400. For works that are not available on CCC please contact mpkbookspermissions@tandf.co.uk

Trademark notice: Product or corporate names may be trademarks or registered trademarks and are used only for identification and explanation without intent to infringe.

Library of Congress Cataloging-in-Publication Data
Names: Thillaiarasu, N, editor. | Tripathi, Suman Lata, editor. |
Dhinakaran, V, editor.
Title: Artificial Intelligence for internet of things : design principle,
modernization, and techniques / edited by N. Thillaiarasu,
Suman Lata Tripathi and V. Dhinakaran.
Description: First edition. | Boca Raton : CRC Press, 2023. |
Series: Smart engineering systems: design and applications |
Includes bibliographical references and index.
Identifiers: LCCN 2022022566 (print) | LCCN 2022022567 (ebook) |
ISBN 9781032210285 (hbk) | ISBN 9781032371986 (pbk) |
ISBN 9781003335801 (ebk)
Subjects: LCSH: Internet of things. | Artificial intelligence.
Classification: LCC TK5105.8857 .A775 2023 (print) |
LCC TK5105.8857 (ebook) | DDC 004.67/8028563–dc23/eng/20221004
LC record available at https://lccn.loc.gov/2022022566
LC ebook record available at https://lccn.loc.gov/2022022567

ISBN: 9781032210285 (hbk)
ISBN: 9781032371986 (pbk)
ISBN: 9781003335801 (ebk)

DOI: 10.1201/9781003335801

Typeset in Times New Roman
by Newgen Publishing UK

Contents

Preface

In the current Industry 4.0 technology revolution, human beings are getting used to various Internet of Things (IoT) devices that include smart watches, smart cars, and smart heart rate sensors, and IoT is becoming one of the most widely used technologies in the world. IoT is employed in many fields such as retail, logistics, pharmaceutical, health care, transportation, housing, urban planning, infrastructure monitoring, aviation, communication networks, earth sciences, disaster management, and the list goes on.

The book titled *Artificial Intelligence for Internet of Things: Design Principle, Modernization, and Techniques* mainly deals with IoT essentials, industrial and medical IoT, machine learning algorithms, robotics, data analytics tools, technologies for smart city, and focuses on the IoT-enabled device design and challenges for real-time system and applications. It also intends to strengthen academic and research activities in the topics like applications and recent trends on IoT technology, transition and transformation towards IoT platform, and IoT edge computing.

The book is organized into 17 chapters. The topics are lucidly explained with examples for the better experience of the reader. The book will cover the fundamental concepts of advanced techniques and tools with the concept of energy-efficient systems. It also covers the upcoming areas of IoT-based automated real-time systems useful in many government and private industries.

This book aims to provide theoretical frameworks and the latest empirical research findings in the IoT area. It is written mainly for the professionals who desire to improve their understanding of the strategic role of trust at various levels of the information and knowledge society, that is, trust at the level of the global economy of networks, organizations and finally trust at the level of individuals in the networked environments.

Editors

N. Thillaiarasu is presently an associate professor in the School of Computing and Information Technology, REVA University, Bengaluru. He has also served as an assistant professor at Galgotias University, Greater Noida from July 2019 to December 2020. He worked from April 2012 to June 2019 as an assistant professor in the Department of Computer Science and Engineering, SNS College of Engineering, Coimbatore. He obtained B.E. in Computer Science and Engineering from Selvam College of Technology in 2010 and received M.E. in Software Engineering from Anna University Regional Center, Coimbatore, in 2012. He received his Ph.D. from Anna University, Chennai, in 2019. He has published more than 25 research papers in refereed journals, and has presented at IEEE Xplore conferences. He has organized several workshops, summer internships, and expert lectures for students. He has worked as a session chair, conference steering committee member, editorial board member, and reviewer in Springer journals (*Journal of Mobile Multimedia*, *Wireless Networks*, *Wireless Personal Communications*, *Cluster Computing*) and IEEE Conferences. He is one of the editors of *Machine Learning Methods for Engineering Application Development* (Bentham Science). He is an editor for the title, *Cyber Security for Modern Engineering Operations Management: Towards Intelligent Industry*, *Artificial Intelligence for Internet of Things: Design Principle, Modernization, and Techniques* (CRC Press, Taylor and Francis). His areas of interest include cloud computing, security, IoT, and machine learning.

Suman Lata Tripathi has completed her Ph.D. in the area of microelectronics and VLSI from MNNIT, Allahabad. She did her M.Tech in Electronics Engineering from UP Technical University, Lucknow and B.Tech in Electrical Engineering from Purvanchal University, Jaunpur. She is associated with Lovely Professional University as a professor and has more than 17 years of experience in academics. She has published more than 55 research papers in refereed IEEE, Springer, and IOP Science journals and conferences. She has organized several workshops, summer internships, and expert lectures for students. She has worked as a session chair, conference steering committee member, editorial board member, and reviewer in international and national and IEEE/Springer journals and conferences.

She received the "Research Excellence Award" in 2019 at Lovely Professional University. She also received the best paper award at IEEE ICICS – 2018. She has edited more than 12 books and one book series in different areas of electronics and electrical engineering. She has edited titles for top publishers like Elsevier, CRC Press, Taylor and Francis, Wiley-IEEE, SP Wiley, Nova Science, and Apple Academic Press. She has edited books titled *Recent Advancement in Electronic Device, Circuit and Materials* (Nova Science Publishers), *Advanced VLSI Design and Testability Issues* and *Electronic Devices and Circuit Design Challenges for IoT Application* (CRC Press, Taylor & Francis and Apple Academic Press). She is also associated as an editor of the book series on *Green Energy: Fundamentals, Concepts, and Applications* and *Design and Development of Energy Efficient Systems* to be published by Scrivener Publishing, Wiley (in production). She is also associated with Wiley-IEEE for her multiauthored (ongoing) book in the area of VLSI design with HDLs. She is also working as series editor for the title, *Smart Engineering Systems* (CRC Press, Taylor and Francis). She has already completed one book with Elsevier on *Electronic Device and Circuits Design Challenges to Implement Biomedical Applications*. She is guest editor of a special issue in *Current Medical Imaging* (Bentham Science). She is a senior member of IEEE, Fellow of IETE and life member ISC and continuously involved in different professional activities along with academic work. Her area of expertise includes microelectronics devices.

V. Dhinakaran is an experienced professor with a demonstrated history of working in higher education and Head at the Centre for Additive Manufacturing in the Chennai Institute of Technology, Chennai, India. Currently he is working as Dean (Research & Development). He has expertise in engineering, finite element analysis (FEA), computational fluid dynamics, composites, welding, additive manufacturing, machine learning, and various other fields. Dhinakaran has given more than 35 expert lectures in various engineering colleges in order to enhance the knowledge of faculty members, researchers, and the student community. He has published 48 SCIE articles and 86 Scopus-indexed articles in peer-reviewed journals. He is the author and editor of various handbooks published by Elsevier, Taylor Francis, and Springer. He is an associate editor of the *Journal of Robotics and Control* and editorial board member of *American Research Journal of Biomedical Engineering* and youth editorial board member of Advanced Powder Materials (APM).

Contributors

S. Adlin Jebakumari
Jain Deemed to be University
Bengaluru, Karnataka, India

M.D. Javeed Ahamed
Narasaraopeta Institute of Technology
Guntur, Andhra Pradesh, India

S.S. Aravinth
Koneru Lakshmaiah Education Foundation
Vaddeswaram, Andhra Pradesh, India

Shaik Bajidvali
Narasaraopeta Institute of Technology
Guntur, Andhra Pradesh, India

Rohini Chavan
Vishwakarma Institute of Information Technology
Pune, Maharashtra, India

Kanupriya Choudhary
Indian Council of Agricultural Research – Central Institute of Agricultural
    Engineering (ICAR–CIAE)
Bhopal, Madhya Pradesh, India

P. Deivendran
Velammal Institute of Technology
Chennai, Tamil Nadu, India

Pallavi Deshpande
Vishwakarma Institute of Information Technology
Pune, Maharashtra, India

Smrity Dwivedi
Indian Institute of Technology
Varanasi, Uttar Pradesh, India

Gauri Ghule
Vishwakarma Institute of Information Technology
Pune, Maharashtra, India

Shubhangi P. Gurway
Priyadarshini Bhagwati College of Engineering
Nagpur, Maharashtra, India

Shraddha Habbu
Vishwakarma Institute of Information Technology
Pune, Maharashtra, India

Essam H. Houssein
Minia University
Minia, Egypt

S. Ilaiyaraja
Velammal Institute of Technology
Chennai, Tamil Nadu, India

Geetanjali Kale
SCTR's Pune Institute of Computer Technology
Pune, Maharashtra, India

S. Kaliappan
Velammal Institute of Technology
Chennai, Tamil Nadu, India

A. Kannagi
Jain Deemed to be University
Bengaluru, Karnataka, India

R. Kavitha
Jain Deemed to be University
Bengaluru, Karnataka, India

Kothalanka Kavyasri
Koneru Lakshmaiah Education Foundation
Vaddeswaram, Andhra Pradesh, India

N. Krishnamoorthy
Kongu Engineering College
Erode, Tamil Nadu, India

Mohit Kumar
Indian Council of Agricultural Research – Indian Agricultural Research Institute
  (ICAR–IARI)
New Delhi, India

Aman Mahore
Indian Council of Agricultural Research – Central Institute of Agricultural
  Engineering (ICAR–CIAE)
Bhopal, Madhya Pradesh, India

Seyedali Mirjalili
Torrens University, Australia

Mohamed AbuBasim
Velammal Institute of Technology
Chennai, Tamil Nadu, India

P. Murugapriya
Surya Engineering College
Erode, Tamil Nadu, India

S. Muthukumar
Velammal Institute of Technology
Chennai, Tamil Nadu, India

Suresh Muthusamy
Kongu Engineering College
Perundurai, Tamil Nadu, India

Rohit Nalawade
Indian Council of Agricultural Research – Central Institute of Agricultural
  Engineering (ICAR–CIAE)
Bhopal, Madhya Pradesh, India

M.S. Nidhya
Jain Deemed to be University
Bengaluru, Karnataka, India

K. Nirmala Devi
Kongu Engineering College
Perundurai, Tamil Nadu, India

Hitesh Panchal
Government Engineering College
Patan, Gujarat, India

Abhishek Patel
Indian Council of Agricultural Research – Central Institute of Agricultural
    Engineering (ICAR–CIAE)
Bhopal, Madhya Pradesh, India

M. Prasad
Vellore Institute of Technology
Chennai, Tamil Nadu, India

P. Radhika Ravi
Senior Industry Principal Independent Consultant and Author
Bengaluru, Karnataka, India

K.S. Raghuram
Vignan's Institute of Information Technology
Visakhapatnam, Andhra Pradesh, India

Ravi Ramaswamy
Principal Consultant – Transformations
Chennai, Tamil Nadu, India

Archana Ratnaparkhi
Vishwakarma Institute of information Technology
Pune, Maharashtra, India

G. Revathy
SASTRA University
Thanjavur, Tamil Nadu, India

D. RAVIKUMAR
Department of ECE, Kings Engineering College
Irugattukotai, Chennai, India

Shasanka Sekhar Rout
GIET University
Gunupur, Odisha, India

Swagat Kumar Samantaray
VIT Bhopal University
Kotri Kalan, Madhya Pradesh, India

S.K. Sanjeev Raja
Kongu Engineering College
Perundurai, Tamil Nadu, India

K. Saravanan
New Horizon College of Engineering
Bengaluru, Karnataka, India

S. Sarumathi
K.S. Rangasamy College of Technology
Namakkal, Tamil Nadu, India

S. Selvakanmani
Velammal Institute of Technology
Chennai, Tamil Nadu, India

K. Selvakumar
Annamalai University
Chidambaram, Tamil Nadu, India

J. Senthilkumar
Sona College of Technology
Salem, Tamil Nadu, India

V. Shanmugasundaram
Velammal Institute of Technology
Chennai, Tamil Nadu, India

S. Sriram
Sri Ramakrishna Polytechnic College
Coimbatore, Tamil Nadu, India

M. Varsha
St. Francis College
Koramangala, Bengaluru, Karnataka, India

P. Velmurugan
Shri Ram Murti Smarak College of Engineering and Technology
Bareilly, Uttar Pradesh, India

K. Velusamy
Sri Vasavi College
Erode, Tamil Nadu, India

K. Venkata Rama Bhagavath
Koneru Lakshmaiah Education Foundation
Vaddeswaram, Andhra Pradesh, India

M. Vignesh
Kongu Engineering College
Perundurai, Tamil Nadu, India

R. Vishnuhari
Kongu Engineering College
Perundurai, Tamil Nadu, India

Kalyani Waghmare
SCTR's Pune Institute of Computer Technology
Pune, Maharashtra, India

Chapter 1

Cyber security control systems for operational technology

S. Sriram

CONTENTS

DOI: 10.1201/9781003335801-1

1.1 INTRODUCTION

Running a business implies you have a ton of resources to secure such as your representatives, cash, and merchandise. The prosperity of your business rides on giving the best insight to your shoppers. In any case, the danger of criminals, hoodlums, and lying clients might put your business in danger (Von Solms and Van Niekerk, 2013; Murray et al., 2017; Brocklehurst et al., 1994; Dacier et al., 1996; Van Brabant, 2000). You need to secure your business and its resources using an incorporated security framework to address every contingency.

Coordinated security frameworks are the most ideal alternatives for any business. In addition to the fact that they offer the best improvements, they will guarantee every one of your bases are covered. Incorporated security frameworks are a sort of complex framework that strings various parts of your business into one. For a long time, modern frameworks depended upon restrictive conventions and programming, were physically overseen and observed by people, and had no association with the rest of the world. They were a genuinely unimportant objective for programmers as there was no organized interface to assault and nothing to acquire. The most ideal way of infiltrating these structures was to get real induction to a terminal and this was no basic task. Information Technology (IT) and Other Technology (OT) incorporated pretty much nothing and didn't manage similar sorts of weaknesses (Murray et al., 2007; Brocklehurst et al., 1994).

Modern frameworks were brought online to convey huge information and shrewd investigation just as they acquire new abilities through innovative incorporations. The union of IT and OT gives associations a solitary perspective on modern frameworks along with measure-the-board arrangements that guarantee that precise data is conveyed to individual network devices, sensors, and gadgets in the best configuration (Sun et al. 2018; Rid and Buchanan, 2015).

IT and OT frameworks work together and their efficiencies have been realized. These frameworks can be distantly observed and the associations can understand security benefits that are utilized on managerial IT frameworks. This progress from shut to open frameworks has produced a huge number of security hazards that should be tended to (Li et al., 2012).

1.2 OPERATIONAL TECHNOLOGY SECURITY RISK

As modern frameworks become more associated, they additionally present weaknesses. The significant expense of modern gear and the destruction to networks and economies that an attack could make are key factors for affiliations wanting to guarantee their modern organizations. Add legacy equipment and security guidelines that might forbid any alterations made to hardware and consistency guidelines that require touchy information to be made accessible to outsiders, and you have a serious test directly. Fortunately, it is feasible to get modern organizations without disturbing activities or encountering rebelliousness (Ullah et al. 2019). By utilizing arrangements that permit total performance of the organization and setting up the right security approaches, you can set up a compelling OT system that will secure your cycles, individuals, and benefits and altogether diminish security weaknesses and occurrences (Van Brabant, 2000).

1.2.1 Today's security of industrial networks

Modern conditions generally have lower signal traffic than IT conditions. A significant part of the traffic signal is going between resolved endpoints and subsequently can be baselined and stocked highly effectively than traffic that is produced in an IT organization. Using checking and examination devices can help with recognizing and guaranteeing against unapproved changes and various irregularities that could signal an attack in full headway or in its fundamental stages. Security is centered around dealing with the insider hazard inside an association and does this through understanding the practices of representatives. With more representatives moving to far-off workplaces, the danger from insiders to an association's touchy information has never been more prominent. These vulnerable sides create the need for a human-driven and robotized way to deal with network safety to assist with securing the individuals and delicate organization information. These dangers have customarily been overseen through a system of setting up fundamental shields to keep dangers out, as opposed to searching for hazard inside an association (Brocklehurst et al. 1994; Dacier et al., 1996).

1.2.2 User activity monitoring

The greatest danger to your touchy information comes from inside user activity monitoring (UAM), which stops dangers the second they're distinguished. Hazardous insiders, regardless of whether malignant or careless, are liable for almost 60 percent of all information penetration. With the mass movement to unaided far-off workplaces, the danger to an association's delicate information has never been more prominent. Any association with restrictive information or protected innovation is defenseless against insider dangers at the human point, where information is generally important and generally helpless. In any case, there are approaches to guard your basic information and customer IP while staying away from substantial fines and loss of income and notoriety. Representatives inside the association can evade security arrangements to cause damage or, all the more regularly, accidentally make moves that compromise even the most note-worthy outer danger protections. There are three threat actor types:

- Compromised – It can allude to innovation (e.g., a particular gadget has been compromised, frequently without the client's full information), or to a person.
- Malicious – Includes both planned demonstrations of treachery for reasons for monetary benefit or reprisal, and purposeful conduct that prompts a break.
- Accidental – Characterized as accidental representative remissness that can uncover your network from messaging client information to some unacceptable beneficiary, to being hoodwinked by the most recent phishing trick.

1.2.3 Hazard in reputed industries

The harms from an information penetration can be huge for any association. Overseeing insider hazard is particularly basic in profoundly managed businesses like healthcare and financial services administrations. Meeting the necessities of administrative norms, including General Data Protection Regulation (GDPR), Health Insurance Portability and Accountability Act (HIPAA), International Organization for Standardization (ISO) 27001, National Institute of Standards and Technology (NIST), Federal Risk and Authorization Management Program (Fed Ramp), Federal Information Security Management Act (FISMA), and others can be a mind-boggling task in the present-day scene. The expanding speed of information availability and an increase in distant labor forces is leaving asso-ciations with vulnerable sides and expected issues with consistency. An insider hazard program ought to keep up with the respectability of information security and protection abilities, while likewise sticking to provincial and industry con-sistency rules. The results of not overseeing insider dangers can be expensive, with occurrences including careless representatives or project workers costing an average loss of $306,110. Insider threat The board enforces consistency while also minimising the risk of fines, legal fees, and reputational damage to the group (Van Brabant, 2000).

1.2.4 Dynamic security battle space

Quite recently, security was not a big threat. However long your organization had have more than normal security, you were possibly safe since another person would get hit first. Programmers were searching for the "easy breaches" – the simple penetrates. Online lawbreakers for the most part utilized a wide "shower and implore" way to deal with sharply discovered targets. Back then, "signature-based" arrangements, which lead to distinguishing known, pernicious program and square them, appeared well and good. In the event that one organization saw another danger, a mark could be composed for it and dispersed to others to shield them from contamination. Quick forward to now. As associations have supported their security, programmers also evolved. Assaults today are more refined and targeted than any other time in recent memory. Maybe rather than sending nonexclusive malware, programmers today cautiously plot every single assault, utilizing special, "zero-day" misuses that render signature-based assurances almost futile.

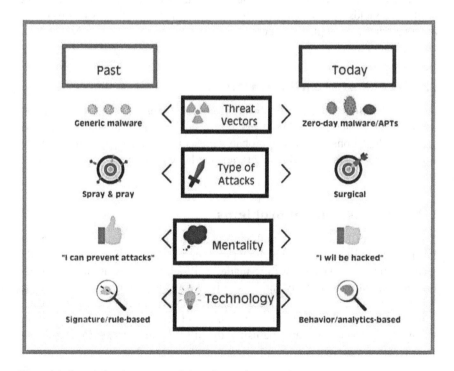

Figure 1.1 Prevention is not enough based on cyber security.

1.3 TAXONOMY OF SECURITY VULNERABILITIES

1.3.1 Buffer overflow

This weakness happens when information is composed past the constraints of a cradle. Cushions are memory regions assigned to a function. By altering the information past the limits of a cushion, the function gets to memory assigned for different cycles. It can prompt a framework crash, data compromise, or give heightening advantages.

1.3.2 Nonsubstantial input

Projects frequently function with information. The data program might be malevolent, intended to constrain a program to act in an accidental manner. For example, the program creates a picture. A noxious client can make a picture document with invalid picture measurements. The noxiously created measurements could constrain the program to allot supports of wrong and startling sizes.

1.3.3 Race conditions

This weakness is the point at which the yield of an occasion relies upon requested or planned yields. A race condition turns into a wellspring of weakness when the necessary arranged or planned occasions don't happen in the right request or legitimate planning.

1.3.4 Lack of security practices

Framework information can be ensured through procedures like confirmation, approval, and encryption. Designers ought not endeavor to make their own security calculations since it will probably present weaknesses. It is firmly understood that engineers use security libraries that have been effectively made, tried, and confirmed.

1.3.5 Access control problems

It is the way of controlling actual admittance to gear directing the approaches to an asset, like a document, and how they can manage it, for example, read the record. Numerous weaknesses are made by ill-advised utilization of controls. Practically, entrance security can be survived if the aggressor has actual admittance to the target . For instance, regardless of whether you set a record's authorizations to, the working framework can't keep somebody from bypassing the working framework and perusing the information circle. This does not ensure data protection.

1.3.6 Malicious software

This program code can be used to modify the data process to induce harm to the system or design structure.

1.3.7 Spyware

This spyware design tracks and watches out for a customer. It oftentimes joins activity trackers, keystroke grouping, and data set. While attempting to overcome well-being endeavors, spyware oftentimes changes security protection.

1.3.8 Program in adware

Promoting upheld programming may be intended to naturally convey notices. It is regularly introduced for certain forms of programming. Some adware is intended to just convey notices yet it is likewise entirely expected for adware to accompany spyware.

1.3.9 Bot

A bot is malware intended to consequently perform activity, normally on the web. A few PCs are contaminated with bots modified to discreetly sit tight for orders given by the attacker.

1.3.10 Ransomware

This malware is installed in the PC framework and holds information as hostage. Ransomware for the most part works by scrambling information in the PC in a way it is obscure to the client. Some different adaptations of ransomware can exploit explicit framework weaknesses to secure the framework. Ransomware is covered by a downloaded record or some product weakness.

1.3.11 Scareware

This malware is intended to convince a system for make a particular move dependent on dread. It manufactures spring up windows that look like working framework exchange windows. These windows pass on manufactured messages expressing the framework is in danger or necessitates the process of program execution to get back to typical activity.

1.3.12 Rootkit

It is a kind of malware intended to adjust the working framework to make a secondary passage. At that point, assailants then utilize the secondary passage to get to the PC distantly. Most rootkits exploit programming weaknesses to perform advantage heightening and adjust framework documents. It is likewise entirely expected for rootkits to alter the framework criminology and observing apparatuses, making them extremely difficult to identify.

1.3.13 Virus

It is an infection of pernicious code connected to documents, frequently in authentic projects. Most infections require end-client enactment and can actuate at a particular time or date. Infections can be innocuous and basically show an image or they can be ruinous, for example, those that alter or erase information. Infections can likewise be customized to transform to keep away from discovery. Most infections are presently spread by peripheral devices or email.

1.3.14 Trojan horse

It is a malware that presents activities under the appearance of an ideal activity. This noxious code misuses the advantages of the client that runs it. Regularly, Trojans are found in picture documents, sound records, or games. It differs from an infection since it ties itself to non-executable records.

1.3.15 Worms

Worms are malicious codes that repeat themselves by autonomously abusing weaknesses in the networks. Worms typically hinder networks. Though an infection requires a host program to run, worms can run without help from anyone else. Other than the underlying contamination, they presently don't need client interest. The worm can spread rapidly over the organization. It share comparative examples. Worms have an empowering weakness, an approach to engender themselves, and they all contain a payload.

1.3.16 Man-in-the-middle [MitM]

It permits the assailant to assume responsibility for gadget except the client's information. With that degree of access, the assailant can catch client data prior to handing off it to its expected objective. MitM assaults are generally used to take monetary data. Numerous malware and procedures exist to give assailants MitM capacities. Despite the kind of framework contaminated with, malware throws the following indications:

- There is an increase in CPU usage and decrease in computer speed.
- The computer freezes or crashes often.
- The content files are modified and deleted.

1.3.17 Blended attacks

Mixed assaults are assaults that utilize numerous methods to think twice about the target. By utilizing a few diverse assault methods immediately, malware are a mixture of worms, Trojan ponies, spyware, keyloggers, spam and phishing plans. This pattern of mixed assaults uncovers more intricate malware and puts client information at an incredible danger.

1.4 METHODOLOGY

1.4.1 Stronger operational technology (OT) security

OT security is the act of utilizing equipment and programming advances to screen, distinguish, and control changes to cycles, occasions, and gadgets. The reason behind utilizing OT security is to secure mechanical frameworks and organizations, for example, savvy city machines and transportation organizations.

1.4.2 Creating inventory and identifying OT vulnerabilities

OT specialists are needed to build up a precise OT resource stock with baselines for each. A total organization map is additionally needed to plan all inbound and outbound communication. A total appraisal ought to be made to distinguish weakness to OT resources and security controls needed to moderate those dangers.

1.4.3 Acquiring automated threat intelligence feeds

Danger insight information gives significant data with respect to dangers. Danger insight takes care of accessible information, including industry, administrative, and business data. Be that as it may, one of the significant concerns is changing over such insight takes care of noteworthy knowledge. To this end, OT security experts ought to send mechanized danger ingestion abilities in network-observing arrangements.

1.4.4 Back/restore

Creating an optimal backup of OT data is a vital approach that can ensure data availability even after the data breach. To this end, first, create a back copy of the OT data and then perform a test restore to ensure that the entire backup system and restore system is working properly and accurately.

1.5 STYLE OF CYBER SECURITY

1.5.1 Security automation

The capacity to consequently coordinate cyber threat intelligence into your security framework for proactive insurance is "Robotized DNS Sinkhole Breach Detection." Security mechanization abilities by locally incorporating cyber threat intelligence into "Computerized Penetration Testing" and "Robotized Incident Response" responds to improved break identification and examination. Sharing cyber threat intelligence, both secretly and openly, would now be able to be performed inside a couple of snaps. Your business and your industry accomplices can now naturally acquire the advantages of industry-explicit cyber threat intelligence through robotized sharing and reconciliation with your security answers for proactive relevant break location and anticipation. It naturally expands the insight of your association to assist with guaranteeing that you keep steady over the most recent dangers, assaults, and security breaks to protect your business.

1.5.2 Breach detection system (BDS)

Break location frameworks classify the uses and security gadgets intended to distinguish the movement of present malware in an organization after a penetrate has happened. Enterprise IT utilizes BDS to secure against the assortment of cutting-edge dangers, particularly unidentified malware. In contrast to level 1 security, like a firewall or interruption anticipation, that checks approaching traffic, BDS centers around malevolent movement inside the organization it secures. It decides potential penetrates by varying the mixes of heuristics, traffic investigation, hazard evaluation, safe stamped traffic, information strategy comprehension, and infringement detailing. Utilizing these techniques, BDS can now and again discover penetrates as they happen and at different occasions distinguish breaks and side-channel assaults that had not recently been found. BDS assists with tracking down the obscure progress and versatile dangers. Indeed, even significant sites have been hacked; moreover, the normal effective break endures for 16 months. On the two checks, there is surely space to eliminate harms. The utilization of BDS involves a change in way of thinking from forestalling each interruption to understanding that interruptions will occur and zeroing in on getting those interruptions sooner. BDS should be designed with subtleties like working framework, a rundown of supported applications, and projects permitted to interface with the Internet.

This load of elements highlight the requirement for hearty break identification to give a "last line of safeguard" against these assaults, rather than simply zeroing in on hindering the underlying influx of the interruption. While afterward location is certainly not another idea, generation of intrusion detection systems (IDS) and security information and event management (SIEM) technologies generally fall short in today's data-center environment. The primary explanation is that they depend on pre-characterized rules/marks to identify penetrates. In a universe

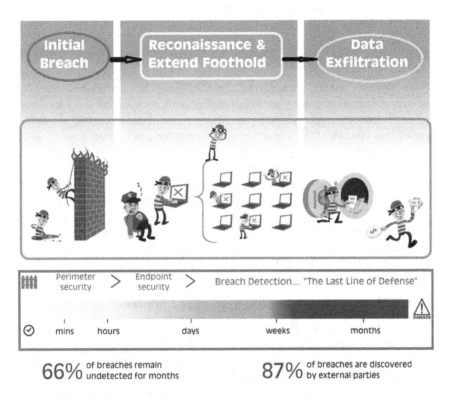

Figure 1.2 Breach detection.

of bespoke assaults, such arrangements will regularly miss the most applicable alarms. It has low sign to-commotion. Envision being overflowed with hundreds or even large number of alarms each day—everyone addressing a likely penetrate to your association's most touchy information. With new penetrate recognition, new businesses can detect these developments and changes. Joined with a cozy comprehension of how programmers work, they can at last sort out all the unique pieces progressively before more critical harm has happened. Eventually, arrangements that adequately slice through the commotion and point up a small bunch of exceptionally applicable, significant security alarms will probably turn into a significant "last line of safeguard" and a vital segment of the cutting-edge venture security stack.

1.5.3 Protection of computing devices from intrusion

1.5.3.1 Keep the firewall-on condition

Regardless of whether it is a product firewall or an equipment firewall on a switch, the firewall ought to be turned on and refreshed to keep programmers from getting to your own or organization information. In Windows 7 and 8.1 or Windows 10, turn on the firewall in the version form of Windows.

1.5.3.2 Antivirus and antispyware

Trojan horses, worms, ransomware, and spyware are introduced on your registering gadgets without your authorization to access your PC and your information. Infections can annihilate your information, hinder your PC, or assume control over your PC. One way infections can assume control over your PC is by permitting spammers to communicate messages utilizing your record. Spyware can screen your online exercises, gather your own data, or produce undesirable spring up advertisements on your internet browser while you are on the web. A decent principle is to just download programming from trusted sites to try not to get spyware in any case. Antivirus programming is intended to check your PC and approaching email for infections and erase them. Here and there, antivirus programming additionally incorporates antispyware. Stay up with the latest to shield your PC from the most up-to-date newest malicious software.

1.5.3.3 Manage your operating system and browser

Programmers are continually attempting to exploit weaknesses in your working frameworks and your internet browsers. To secure your PC and your information, set the security settings on your PC and program at medium or higher. Update your PC's working framework including your internet browsers and consistently download and introduce the most recent programming patches and security refreshes from the sellers.

1.5.3.4 Protection of smart devices

Your registering gadgets, regardless of whether they are PCs, workstations, tablets, or advanced cells, ought to be password secured to forestall unapproved access. The put away data ought to be scrambled, particularly for touchy or private information. For cell phones, just store essential data, in the event that these gadgets are taken or lost when you are away from your home. On the off chance that any of your gadgets is compromised, the lawbreakers might approach every one of your information through your distributed storage specialist organization, for example, iCloud or Google Drive.

1.5.3.5 Unique passwords for each online account

Most likely have more than one online record, and each record ought to have a special secret phrase. That is a great deal of passwords to recollect. The outcome of not utilizing solid and exceptional passwords leaves you and your information powerless against digital lawbreakers. Utilizing similar secret word for all your online records resembles utilizing similar key for all your locked entryways. If an aggressor was to get your key, he would get to all that you own. On the off chance that lawbreakers get your secret key through phishing, for instance, they will attempt to get into your other online records. In the event that you just utilize one secret phrase for all records, they can get into every one of your records, take or delete every one of your information, or choose to mimic you. Online records that need passwords that is in turn requires excessive memory to recollect. One answer for abstaining from reusing passwords or utilizing powerless passwords is to utilize a secret word supervisor. A secret word chief stores and encodes the entirety of your unique and complex passwords. The administrator would then be able to assist you with signing into your online records consequently. You just need to recollect your lord secret key to get to the secret phrase director and deal with the entirety of your records and passwords.

1.5.3.6 Detecting attacks in real time

Programming isn't awesome. At the point when a programmer abuses a blemish in a piece of programming before the maker can fix it, it is known as a zero-day assault. Because of the complexity and immensity of zero-day assaults discovered today, it is turning out to be entirely expected that organization assaults will succeed and that a fruitful safeguard is currently estimated in how rapidly an organization can react to an assault. The capacity to identify assaults as they occur continuously, just as halting the assaults promptly, or not long after happening, is the best objective. Lamentably, numerous organizations and associations today can't identify assaults until days or even a very long time after they have happened. Constant scanning from edge to endpoint has to be done – detecting assaults progressively requires effectively examining for assaults utilizing firewall and IDS/IPS network gadgets. Cutting-edge customer/worker malware discovery with associations with online worldwide danger communities should likewise be utilized. Today, dynamic examining gadgets and programming should distinguish network inconsistencies utilizing setting-based investigation and conduct recognition.

1.5.3.7 Cyber attacks in operational technology

Table 1.1 Cyber attacks in 2020

Cyber attacks	Organization
Software AG affected by a ransomware attack in October 2020	Software AG, Germany
Sopra Steria French IT service hit by ransomware attack	Sopra Steria, French IT service giant
Seyfarth Shaw LLP hit by an "aggressive malware" attack	Seyfarth Shaw LLP, global legal firm
Carnival Corporation reported a data breach hit by ransomware attack	Carnival Corporation, cruise line operator

1.6 AVOIDANCE OF THREADS IN OPERATIONAL TECHNOLOGY

1.6.1 Distributed denial of services (DDoS) attacks and response

DDoS is one of the greatest assault dangers requiring constant reaction and location. DDoS assaults are amazingly hard to shield against on the grounds that the assaults begin from hundreds, or thousands of zombies, and the assaults show up as real traffic, as displayed in the figure. For some organizations and associations, consistently happening DDoS assaults cripple internet workers and organization accessibility. The capacity to recognize and react to DDoS assaults continuously is vital.

1.6.2 Protecting against malware in operational technology

Advanced malware protection (AMP) threat grid investigates a large number of records and relates them against countless other dissected malware antiquities. This gives a worldwide perspective on malware assaults, crusades, and their dispersion. AMP is customer/worker programming conveyed on endpoints, as an independent worker, or on other organization security gadgets.

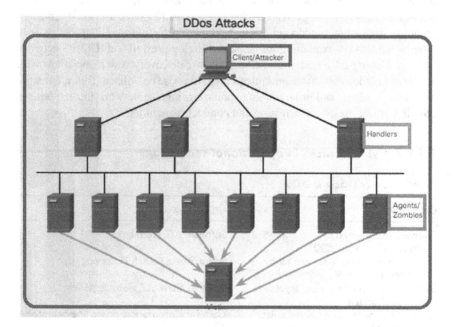

Figure 1.3 Distributed denial of services (DDoS) attacks.

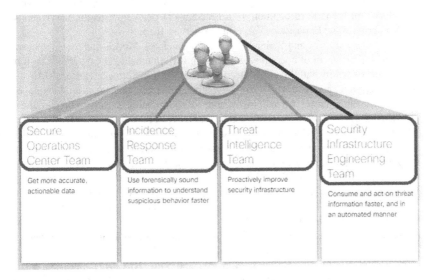

Secure Operations Center Team
Get more accurate, actionable data

Incidence Response Team
Use forensically sound information to understand suspicious behavior faster

Threat Intelligence Team
Proactively improve security infrastructure

Security Infrastructure Engineering Team
Consume and act on threat information faster, and in an automated manner

Figure 1.4 Advanced malware protection (AMP) threat grid.

1.7 CONCLUSION

Operational technology (OT) frameworks are generally being utilized in modern conditions and network protection has become their fundamental issue. Nonetheless, you can keep away from and moderate OT dangers by proactively recognizing, characterizing, and checking your OT foundation. To proactively identify, you are required to use a threat hunting technique. You can use an automated threat hunting system. The safety efforts ought to be planned and implemented. Initially, an organization should know its need for safety on the various levels of the association and afterward it ought to be carried out for various levels. Security approaches ought to be planned first before its execution in such a manner that future adjustment and selection can be adequate and effectively reasonable. The security framework should be tight. However, it should be adaptable for the end-client..

Bibliography

Brocklehurst, S., Littlewood, B., Olovsson, T., and Jonsson, E. (1994). On measurement of operational security. *IEEE Aerospace and Electronic Systems Magazine*, 9(10), 7–16.

Cashell, B., Jackson, W. D., Jickling, M., and Webel, B. (2004). The Economic Impact of Cyber-attacks. Congressional Research Service Documents, CRS RL32331, Washington DC, p. 2.

Cavelty, M. D. (2010). *Cyber-security: The Routledge Handbook of New Security Studies* (pp. 154–162). London: Routledge.

Cheung, S., Lindqvist, U., and Fong, M. W. (2003, April). Modeling multistep cyber attacks for scenario recognition. In Proceedings DARPA Information Survivability Conference And Exposition (Vol. 1, pp. 284–292). IEEE.

Dacier, M., Deswarte, Y., and Kaâniche, M. (1996, January). Models and tools for quantitative assessment of operational security. In IFIP International Conference on ICT Systems Security and Privacy Protection (pp. 177–186). Springer, Boston, MA.

Goo, J., Yim, M. S., and Kim, D. J. (2014). A path to successful management of employee security compliance: An empirical study of information security climate. *IEEE Transactions on Professional Communication*, 57(4), 286–308.

Haddad, S., Dubus, S., Hecker, A., Kanstrén, T., Marquet, B., and Savola, R. (2011, September). Operational security assurance evaluation in open infrastructures. In 2011 6th International Conference on Risks and Security of Internet and Systems (CRiSIS) (pp. 1–6). IEEE.

Hideshima, Y., and Koike, H. (2006, January). STARMINE: A visualization system for cyber attacks. In Proceedings of the 2006 Asia-Pacific Symposium on Information Visualisation, Volume 60 (pp. 131–138).

Julisch, K. (2008, September). Security compliance: the next frontier in security research. In Proceedings of the 2008 New Security Paradigms Workshop (pp. 71–74).

Li, X., Liang, X., Lu, R., Shen, X., Lin, X., and Zhu, H. (2012). Securing smart grid: cyber attacks, countermeasures, and challenges. *IEEE Communications Magazine*, 50(8), 38–45.

Murray, G., Johnstone, M. N., and Valli, C. (2017). *The Convergence of IT and OT in Critical Infrastructure*. https://ro.ecu.edu.au/cgi/viewcontent.cgi?article=1217&context=ism

Rid, T., and Buchanan, B. (2015). Attributing cyber attacks. *Journal of Strategic Studies*, 38(1–2), 4–37.

Safa, N. S., Von Solms, R., and Furnell, S. (2016). Information security policy compliance model in organizations. Computers and Security, 56, 70–82.

Sun, C. C., Hahn, A., and Liu, C. C. (2018). Cyber security of a power grid: State-of-the-art. *International Journal of Electrical Power and Energy Systems*, 99, 45–56.

Thakur, K., Qiu, M., Gai, K., and Ali, M. L. (2015, November). An investigation on cyber security threats and security models. In 2015 IEEE 2nd International Conference on Cyber Security and Cloud Computing (pp. 307–311). IEEE.

Ullah, F., Naeem, H., Jabbar, S., Khalid, S., Latif, M. A., Al-Turjman, F., and Mostarda, L. (2019). Cyber security threats detection in internet of things using deep learning approach. *IEEE Access*, 7, 124379–124389.

Uma, M., and Padmavathi, G. (2013). A survey on various cyber attacks and their classification. *IJ Network Security*, 15(5), 390–396.

Van Brabant, K. (2000). *Operational Security Management in Violent Environments* (p. 330). London: Overseas Development Institute.

Von Solms, R., and Van Niekerk, J. (2013). From information security to cyber security. *Computers and Security*, 38, 97–102.

Zhu, B., Joseph, A., and Sastry, S. (2011, October). A taxonomy of cyber attacks on SCADA systems. In 2011 International Conference on Internet of Things and 4th International Conference on Cyber, Physical and Social Computing (pp. 380–388). IEEE.

Chapter 2

Cyber security ecosystem and essentials for modern technology world

*S.S. Aravinth, K. Venkata Rama Bhagavath,
Kothalanka Kavyasri, and J. Senthilkumar*

CONTENTS

DOI: 10.1201/9781003335801-2

2.1 INTRODUCTION

Unlike physical security, this mechanism protects our digital assets in terms of digital information stored in a variety of electronic mediums. The intruders are taking control of our digitally stored data. To protect, it is essential to apply and implement a security mechanism. The traditional methods and techniques such as encryption and decryption are used globally with the cyber security mechanisms to provide a higher degree of protection.

Cyber security involves the technologies, processes and controls designed to protect the systems, networks, programs, devices and data from cyber attacks.

Figure 2.1 explains the major focus of cyber security principles as integrity, confidentiality and availability.

2.1.1 Confidentiality

Confidentiality is considered as a privacy and restricted access of the information from an unauthorized agent. With the blocking of access, this confidentiality of the information is maintained.

Figure 2.1 The triad of cyber security mechanism.

2.1.2 Integrity

It is a process of preserving data after the information access has been granted to the subscribed users. It also focuses more on maintaining the recovery of data and backup of already existing data.

2.1.3 Availability

The request and response model is applied to access the data from the user to service provider. Because of this, the hierarchy of data extraction and process would be taken care in a great manner.

Apart from the above-mentioned principles, some secondary terms are also gaining attention after the emergence of social media. This is a two-phase authentication of user identity.

2.2 CYBER SECURITY LAYERS, SECURING OS, ANTIVIRUSES AND MALWARES

2.2.1 Cyber security layers

Cybersecurity is a monolithic piece of technology that improves security. It is supposed to be a layered approach with multiple facets to ensure system security.

The seven layers are as follows:

1. Mission-critical assets
   This is the first layer, which is faster in technology, that detects malware in your systems by masks and hashes.
   An example for mission-critical assets is electronic medical record in the health care industry.
2. Data security
   It is mainly focused on protecting the data when it is being transferred and stored in your system. The backup security should be present for preventing the loss of your data and this will require the usage of encryption and decryption for the data.
3. Endpoint security
   It provides protection on a network depending on the needs of a business. This security layer makes sure that the endpoints of user devices are not exploited by breaches. This includes the protection of mobile devices, desktops and laptops.
4. Application security
   This application layer provides the security that controls access for the application. It also includes the internal security of the app itself. Mostly, the applications are provided with enough security when they are being used.

5. Network security
    Network security is a set of rules and configurations that are designed to protect the integrity, confidentiality and accessibility of interconnected networks and data that are being used by software platforms and hardware interfaces.
6. Perimeter security
    The main focus of this layer is on both physical and digital security methods to protect the business-level network. Firewalls are being used for securing the data against external forces.
7. Human security
    We all know that the human layer is the weakest in the security chain. It is more prone to psychological manipulation. Here we aim to protect it since it deals with business logic and emotional behaviour, among other things.

2.2.2 Securing OS

Operating system (OS) security is essential for all aspects for maintaining system health, performance since OS is key to any kind of operations that runs on the system.

The potential threats for the operating systems should be eliminated at any cost. Either it can be viruses or trojans.

Well then how?

By authentication.

It is advised that we are supposed to ensure that we are having an under-privileged user since if any attack happens, the root will be secured under this layer.

User authentication can be done either by a one-time password (OTP), password or biometrics.

2.2.3 Antiviruses

Antivirus is a computer program which detects and warns about any unwanted or unrecognised threats that have been running in the device and have the potential to exploit the system.

2.2.4 Malwares

Malware is a computer program which is unrecognised by the user by any means. Typically, malware is installed by the attacker in the victim's device either to monitor the activities or to exploit sensitive information.

Figure 2.2 clearly explains the differences between antivirus software versus malwares.

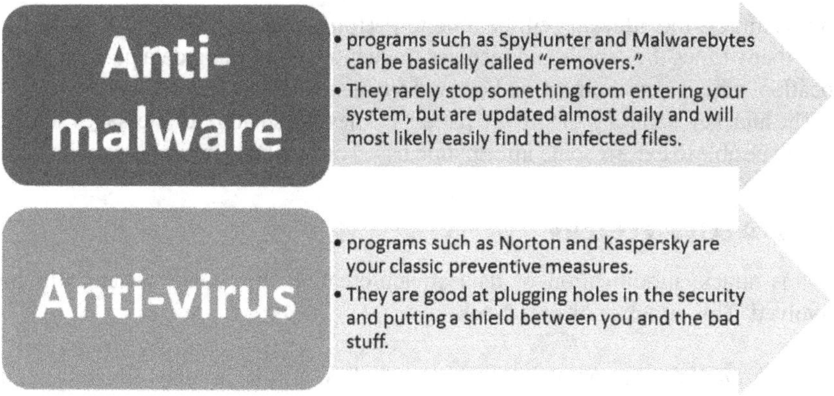

Anti-malware

- programs such as SpyHunter and Malwarebytes can be basically called "removers."
- They rarely stop something from entering your system, but are updated almost daily and will most likely easily find the infected files.

Anti-virus

- programs such as Norton and Kaspersky are your classic preventive measures.
- They are good at plugging holes in the security and putting a shield between you and the bad stuff.

Figure 2.2 Antiviruses versus malwares.

2.3 CYBER ATTACKS AND STRATEGIES

2.3.1 Security attacks

Someone comprised the security level of the system (desktop and laptop) for a particular individual/organisation to steal a confidential information from them. There are two types of attacks: (1) passive attack and (2) active attack.

2.3.1.1 Passive attack (reading of messages)

This attack is mainly performed to gain access to information from the system.

1. Generally, it is an unauthorised reading of messages. It doesn't affect system resources.
2. In this attack, attackers are able to see exchange information between the sender and the receiver, but no modifications of information will take place by the attacker.

2.3.1.2 Passive attack

1. Release of message content
2. Traffic analysis

2.3.1.2.1 Release of message content

In this attack, if two persons are exchanging confidential information through internet/any communicating devices, then the third person (attacker) may steal that information and he will be able to read that information.

2.3.1.2.2 Traffic analysis

In this attack, the attacker will be able to perform packet sniffing, which means how many times the sender and receiver frequently are sending data packets, their location, who is communicating host and frequency of messages.

The attacker is aiming to find the pattern of messages/nature of messages so that he will be able to extract some information based on available information with him.

2.3.2 Active attacks

In this attack, modification of data stream/creation of a false system will be involved. This attack is of four types.

1. *Masquerade*: An authorized entity with few privileges may demand extra privileges.
2. *Replay*: In this attack, the attacker will be capture the message from the sender, later modify that message and send that modified message to the receiver.
3. *Modification of message*: In this attack, the attacker will modify original message with his existing message and send that modified message.
4. Denial of service (DOS): The attacker disrupts the service provided by the server.

2.3.3 Cyber security strategies

2.3.3.1 Creating a secure cyber ecosystem

It involves three symbiotic structures – automation, interoperability, and authentication.

Automation: It simplifies the execution of advanced security measures that ensures the following things:

1. Speed is increased.
2. Develop the decision-making processes.

Interoperability: It hardens collaborative actions, increases awareness and speeds up the learning process.

Authentication: It develops the identification and verification technologies that work to provide

- Security
- Affordability
- Ease of use and administration
- Scalability
- Interoperability.

2.4 NETWORK SECURITY

We know that "a group of computers are connected either to establish connection for communication or sharing of the resources is called a computer network."

Then what is network security? Let's discuss it now.

It comprises of the policies and practices implemented to prevent and monitor unauthorized access or misuse of the sensitive information that is needed to be protected.

"Ok now, we understood what is need to be done, and now we'll get to know how to achieve our goal."

2.4.1 Network components

When we are dealing with the networks, we are supposed to have a thorough understanding. Networks have both physical and software components which are responsible for the transfer of the data or sharing the resources.

NIC, modem, switch, cable, router, and hubs will fall under the physical requirements of the networking and protocols, and operating systems are software requirements of the network.

- Hardware components
- *NIC*: NIC stands for network interface card, which is used to connect the computers in a network and responsible for data preparation that is going to be transferred from one system to another system.

  NIC has two variations: wired NIC and wireless NIC.

  Wired NIC would be suitable for the connections that use cables to transfer the data.

  Wireless NIC has an antenna to establish the wireless connection over the devices. Most of the modern devices have wireless NICs.
- *Hub*: A hub is a network that is shared between the multiple devices is divided by the hub. It searches for the device in the network when the request is raised by the device within that network. If not found, the request would get dropped.

  Hubs became extinct after the introduction of switch routers and modems since they use relatively less bandwidth than hubs to communicate.
- *Switch*: Switches are more advanced version than the hubs. It has an updatable table that continuously iterates the data status whether it is transmitted to the destination or not.

  Unlike hub, the data is not broadcasted. It is just transmitted across the network to the required destination.
- *Router*: Routers and switches provide service for the same layer of OSI model. Unlike switches permitted only for the devices, routers can connect to multiple switches and can also provide the internet facility to the users.

- *Modems*: Modems allow the PC to connect internet over the telephone line. Modem stands for modulation and demodulation since they convert the digital signals to analog signals.
  Software components
- *Operating systems*: Operating systems are essential part of the networks, since every system in the network obviously has an operating system guiding the programs in the system to run the network properly.
- *Protocols suite*: Protocols are set of instructions that a network supposed to follow.

For example, OSI, TCP/IP, and HTTP.

2.4.2 Securing networks

The users and network administrators are advised to implement the following recommendations to better secure their network infrastructure:

- Segmentation and segregate network and functions
  - This helps intruder from propagating in small networks. Based on the access parameters like user role and their functionalities, it ensures there is no propagation of the exploits.
- Limit unnecessary lateral communications.
  - Disabling unfiltered peer-to-peer communications over the large networks prevents the exploitation to stop spreading from multiple pool of systems.
- Hardening and validation of network devices.
  - Ensuring better encryption strategies are applied in network devices and testing them periodically would save a lot of data from getting breached. Also restricting the physical access to switches and routers lowers the chance of getting exploited.
- Performing out-of-band management.
  - OoB management removes the user traffic from the network equation. It uses the alternate communication path to analyse the devices in infrastructure.
  - It is expensive to maintain although it can be installed either by physical, virtual or by hybrid modes.

2.5 CYBER SECURITY NETWORK AND POLICIES

What is network policy? A network security policy delineates guidelines for computer network access, determines policy enforcement and lays out the architecture of the organization's network security environment and defines how the security policies are implemented throughout the network architecture.

Network security policies describe an organization's security controls. It aims to keep malicious users out while also mitigating risky users within your organization. The initial stage to generate a policy is to understand what information and services are available, and to whom, what the potential is for damage and what protections are already in place.

The security policy should define the policies that will be enforced. This is done by dictating a hierarchy of access permissions – granting users access to only what they need to do their work.

These policies need to be implemented in your organization's written security policies and also in your IT infrastructure – your firewall and network control security policies.

What is the purpose of cyber security policy? Cybercrimes are becoming more and more common across the world, making cyber security of the top priorities for everyone. Consequently, there has been a rapid increase in various cyber laws.

In order to protect your company from numerous cybercrimes, you should have a clear and organized cyber security company policy.

The following data security systems in a company would possibly need a lot of attention in terms of security:

- Encryption mechanisms – antivirus systems
- Access control devices – websites
- Authentication systems – gateways
- Firewalls – Routers and switches

The given below are the goals of security policies:

- To maintain an outline for the management and administration of network security
- To protect an organization's computing resources
- To eliminate legal liabilities arising from workers or third parties
  - To prevent wastage of company's computing resources
  - To prevent unauthorized modifications of the data
  - To scale back risks caused by illegal use of the system resource
  - To differentiate the user's access rights
  - To protect confidential, proprietary data from theft, misuse, and unauthorized disclosure.

2.5.1 Types of security policies

A security policy is a document that contains data about the way the company plans to protect its data assets from known and unknown threats. These policies help to keep up the confidentially, availability and integrity of data. The four major forms of security policy are as following.

2.5.1.1 Promiscuous policy

This policy doesn't enforce any restricted access of system resources. For example, with a promiscuous net policy, there's no restriction on net access. A user will access any web site, transfer any application and access a laptop or a network from a foreign location. This may be helpful for businesses wherever people travel or work branch stations need to access the structure networks. Several malwares, virus and Trojan threats are residing on the internet and because of free net access, this malware will return as attachments but not the data of the user. Network directors should be very alert in selecting this kind of policy.

2.5.1.2 Permissive policy

Policy begins undecided and the known harmful activities are blocked. For instance, in a very permissive net policy, the bulk of net traffic is accepted. However, many proverbial dangerous services and attacks square measure blocked. As a result of blocking solely proverbial attacks and exploits, it's not possible for directors to stay up with current exploits. Directors are perpetually enjoying catch-up with new attacks and exploits. This policy ought to be updated often to be effective.

2.5.1.3 Prudent policy

A prudent policy starts with all the services blocked. The administrator permits safe and necessary services singly. It logs everything, like system and network activities. It provides most security whereas permitting only proverbial, however, necessary dangers.

2.5.1.4 Paranoid policy

A paranoid policy forbids everything. There's a strict restriction on all use of company computers, whether or not it's system usage or network usage. There's either no net association or severely restricted net usage. Because of these to a fault severe restrictions, users typically try and notice ways to get around them.

2.5.2 Examples of security policies

Given below square measure samples of security policies that organizations use worldwide to secure their assets and vital resources.

2.5.2.1 Access management policy

Access management policy outlines procedures that facilitate in protecting the structure resources and also the rules that management access to them. It permits organizations to trace their sets.

2.5.2.2 Remote-access policy

A remote-access policy contains a collection of rules that define authorized connections. It defines who will have remote access, the access medium and remote-access security controls. This policy is critical in larger organizations during which networks are geographically dispersed, and during which employees work from home.

2.5.2.3 Firewall management policy

A firewall management policy defines a standard to handle application traffic, like net or e-mail. This policy describes the way to manage, monitor, protect, and update firewalls within the organization. It identifies network applications, vulnerabilities related to applications and creates an application–traffic matrix showing protection strategies.

2.5.2.4 Network connection policy

A network connection policy defines the set of rules to secure network connectivity, including standards for configuring and extending any part of the network, policies related to private networks and detailed information about the devices attached to the network. It protects against unauthorized and unprotected connections that allow hackers to enter into the organization's network and affect data integrity and system integrity. It permits only authorized persons and devices to connect to the network and defines who can install new resources on the network, as well as approve the installation of new devices, and document network changes, etc.

2.6 DISASTER RECOVERY PLANNING

2.6.1 What is disaster recovery planning (DRR)?

Disaster recovery plans (DRP) seek to quickly redirect available resources into restoring data and information systems following a disaster. A disaster can be classified as a sudden event, including an accident or natural disaster, that creates wide scoping, detrimental damage. In information management, DRPs are considered a critical subset of an entity's larger business continuity plan (BCP), which seeks to prepare for, prevent, and recover from potential threats affecting an organization. While BCPs address all facets of an organization, DRPs specifically focus on technology. DRPs provide instructions to follow when responding to various disasters, including both cyber and environment-related events. DRPs differ from incident response plans that focus on information gathering and coordinated decision-making to understand and address a specific event.

2.6.2 Why does it matter?

When DRPs are properly designed and executed, they enable the efficient recovery of critical systems and help an organization avoid further damage to mission-critical operations. Benefits include minimizing recovery time and possible delays, preventing potential legal liability, improving security and avoiding potentially damaging last-minute decision-making during a disaster.

Apart from their specific focus on technology, DRPs and the process for developing them are no different than the range of emergency response protocols and backup plans that election officials have already developed to address potential issues or disruptions. The lessons learned from those exercises are often valuable to DRP development. Election officials develop these plans due to the potential risk impacts during key operational periods, such as the last day for voter registration or candidate filing and election day. For example, if all voting machines were damaged during a flood while in storage just before an election, having an effective DRP could minimize the impact and reduce recovery time.

2.6.3 What you can do?

Election offices should have a comprehensive DRP in place and regularly exercise it to ensure effectiveness. The US Election Assistance Commission has published helpful tips for contingency and disaster recovery planning that election offices can leverage during this process. In order to create an effective DRP, the EI-ISAC recommends:

- including relevant stakeholders from the various business units that may be impacted in the planning process,
- conducting a business impact analysis (BIA) to identify and prioritize critical systems,
- exercising the DRP to test its efficacy,
- conducting after action reviews to identify what went right, what went wrong, and annotate improvements, and
- regularly reviewing the DRP to ensure contacts are up to date and procedures are still effective and relevant.

Election offices should also consider personnel training in the specifics of disaster recovery planning or leverage third-party resources for the planning and recovery process. The MS-ISAC Business Resiliency Workgroup has resources available upon request to assist election officials in creating, testing and improving their DRPs, including a BIA guide and template, suggested items to include in a go-bag, exercise scenarios and an After Action Report template.

2.7 CONCLUSION

As we saw, taking a few simple steps can go a long way toward protecting your privacy and making your laptop more secure to make things even easier, I've created an infographic that summarizes them.

If you find this advice useful, don't forget to share this post or the infographic on your favourite social networks so your friends and colleagues can also better protect their laptops.

To be notified when my next post is online, follow me on Twitter, Facebook, or LinkedIn. You can also get the full posts directly in your inbox by subscribing to the mailing list or the RSS feed.

Bibliography

Ahmed, J., Tushar, Q. "COVID-19 pandemic: a new era of cyber security threat and holistic approach to overcome," 2020 IEEE Asia-Pacific Conference on Computer Science and Data Engineering (CSDE), 2020, pp. 1–5, doi: 10.1109/CSDE50874.2020.9411533.

Chen, Z. "Deep learning for cybersecurity: a review," 2020 International Conference on Computing and Data Science (CDS), 2020, pp. 7–18, doi: 10.1109/CDS49703.2020.00009.

Ghosal, S., Sengupta, S., Majumder, M., Sinhad, B. "Intrusion detection in cyber security: Role of machine learning and data mining in cyber security", *Advances in Science, Technology and Engineering Systems*, Volume 5, Issue 3, 2020, pp. 72–81.

Muzammil Parvez, M., Ravindran, R.S.E., Inthiyaz, S., Tejkumar, C., Veera Ram Sai, K., Shiva Reddy, K.A., "Network security using notable cryptographic algorithm for IoT data", *International Journal of Emerging Trends in Engineering Research*, Volume 8, Issue 5, 2020, Article number 111, pp. 2169–2172.

Sai, J.A., Kumar, K.K., "Enabling public auditing for data integrity checking using ring signatures developed with AES in cyber security", *International Journal of Innovative Technology and Exploring Engineering*, Volume 8, Issue 6, April 2019, pp. 1536–1543.

Sajal, S.Z., Jahan, I., Nygard, K.E. "A survey on cyber security threats and challenges in modem society," 2019 IEEE International Conference on Electro Information Technology (EIT), 2019, pp. 525–528, doi: 10.1109/EIT.2019.8833829.

Suguna, Y.S., Reddy, B.K., Durga, V.K., Roshini, A., "Secure Quantum Key distribution encryption method for efficient data communication in wireless body area sensor net-works", *International Journal of Engineering and Technology(UAE)*, Volume 7, Issue 2.32 Special Issue 32, 2018, pp. 331–335.

Swetha, K., Narasinga Rao, M.R., "Dynamic searchable encryption over distributed cloud storage", *Asian Journal of Information Technology*, Volume 15, Issue 23, 2016, pp. 4763–4769.

Tamma, L.N.D., Ahamad, S.S., "A novel chaotic hash-based attribute-based encryption and decryption on cloud computing", *International Journal of Electronic Security and Digital Forensics*, Volume 10, Issue 1, 2018, pp. 1–19.

Thakur, K., Qiu, M., Gai, K., Ali, M.L., "An investigation on cyber security threats and security models," 2015 IEEE 2nd International Conference on Cyber Security and Cloud Computing, 2015, pp. 307–311, doi: 10.1109/CSCloud.2015.71.

Veerapaneni, S.S., Sekhar, K.R., "A systematic study of asset management using hybrid cyber security maturity model", *International Journal of Recent Technology and Engineering*, Volume 7, Issue 6, April 2019, pp. 140–145.

Chapter 3

Smart manufacturing in Industry 4.0 using computational intelligence

G. Revathy, K. Selvakumar, P. Murugapriya and D. Ravikumar

CONTENTS

3.1 INTRODUCTION

The foundation of Industry 4.0 is the expansion of data sets that are available. Existing traditional technologies are incapable of processing various sorts of available, vast, and complicated data collections, referred to as big data. In order to capture and extract data from the industrial environment [1–4], advanced methods, technologies, algorithms, and software must be applied. Machine learning (ML) [4, 5] is the field that encompasses all of these disciplines. Artificial intelligence (AI) is rapidly becoming the most essential tool for predicting and classifying the difficulty of addressing problems in production systems [5]. AI not only makes

DOI: 10.1201/9781003335801-3

use of greater processing power and diverse applications to quorate convenient material and acquaintance from data acquired from the environment, but it also has the potential to learn from that data using artificial/computational intelligence [5,6]. AI can attain a higher level of criteria than humans for some specific tasks. This emphasises the significance of the data used to obtain the information. However, a balance must be struck. Another significant issue is the data security aspect [7]. With that in mind, AI must employ a variety of strategies and algorithms in order to get the most out of the data [4,6]. The most common learning approaches, as classified by the feedback available, are supervised, unsupervised, and reinforcement learning methods [8]. The problems and applications of AI in today's manufacturing systems are the subject of this chapter. In addition, future AI developments in industrial applications are highlighted, with the primary goal of data exploitation to achieve cost-effective, fault-free, and optimal quality manufacturing processes. A change in manufacturing processes towards supply chain integration, as well as the integration of physical and cyber capabilities, is referred to as smart manufacturing (SM). It's sometimes confused with "Industry 4.0," a term coined by To construct a smart factory, the German government is promoting a fourth generation of manufacturing that encompasses concepts like cyber-physical systems, virtual representations of real equipment and processes, and decentralised decision making [3–5]. All sectors are developing these skills, although at various rates, based on factors such as installed base, culture, supplier base, and demand. The next part goes over the history of factory operations analytics, the rise of SM, and the future of big data analytics applications in the semiconductor industry. In semiconductor manufacturing, the issues that face analytics development in the manufacturing industry are described, as well as a taxonomy for understanding analytics capabilities, the present state-of-the-art, and an analytics roadmap for supporting SM principles. The case study applications that follow show the potential of analytics approaches as well as the SM trends that they foster.

3.2 ARTIFICIAL INTELLIGENCE

The following decades had ups and downs, with two major winters in 1974–1980 and 1987–1993 [4]. Symbolic and sub-symbolic AI approaches are the two categories of AI approaches (e.g., neural networks, fuzzy sets, and evolutionary algorithms). Symbolic intelligence had a lot of success in the first wave, in the 1960s, imitating high-level thinking in small demo programmes for solving "toy problems." Meanwhile, neural network and cybernetics-based techniques have been abandoned or put to the background [5].

The abandoning of connectionism in 1969, when Minsky and Papert published *Perceptrons*, a book outlining the limitations of perceptrons [6], as well as a significant drop in AI research. In reaction to the Lighthill Report of 1973 and early 1970s DARPA budget cuts, AI suffered through its first winter in 1974–1980. Expert systems (ESs) and knowledge-based systems (KBSs) were popular in the

1980s, with knowledge bases comprising high-level subject information gathered from experts and expressed logically.

Manufacturing [7] has also created apps to handle tough machine programming problems that require many tools to work on the same workpiece at the same time. Knowledge engineering/knowledge representation, which is crucial in classic AI research, arose as a result of this necessity. Meanwhile, the "winter" of Hopfield, Rumelhart, and other connectionists reignited worldwide interest in neural networks. As a result, academics began to pay attention to sub-symbolic techniques such as neural networks (NNs), fuzzy sets, statistics, and evolutionary algorithms, which did not require specialised knowledge representations (genetic algorithm (GA)).

As a result, AI systems began to employ sub-symbolic techniques. The loss of the Lisp machine market in 1987, the Strategic Computing Initiative's withdrawal of fresh AI expenditure in 1988, the failure of fifth-generation computers, and the fall of expert systems in the early 1990s all contributed to the second AI winter [4]. Distributed AI (DAI) became more popular in the 1990s. as well as the transformation of data into knowledge [8]. The current surge in AI interest began in 2010 and was spurred by three interconnected factors [9].

Machine learning approaches and algorithms have improved dramatically as a result of the raw data provided by big data, and powerful computers that support big data computing. Big data comes from a variety of places, including e-commerce, social media, the research community, organisations, and government; machine learning approaches and algorithms have improved dramatically as a result of the raw data provided by big data; and powerful computers that support big data computing.

Traditional AI (known as AI 1.0), which emphasised symbolic approaches with structured contents and centralised control structures, is being displaced by Artificial Intelligence 2.0 (AI 2.0), which emphasises machine learning (especially deep learning) with unstructured contents and decentralised (distributed) control structures. Figure 3.1 displays the evolution of AI from the perspective of content and control, from AI 1.0 (Symbolic AI) to AI 2.0 via DAI (Distributed AI) or Web AI, from AI 1.0 (Symbolic AI) to AI 2.0 via DAI (Distributed AI) or Web AI (AI 1.5D or AI 1.5W). Subsymbolic techniques have become more prevalent in the AI evolution process.

Symbolic techniques dominated AI at the time, despite the reappearance of neural networks in the 1980s. Hybrid expert systems were later created by combining traditional expert experience, artificial neural networks, evolutionary algorithms, and fuzzy sets in various combinations. By using training patterns rather than loading rules, neural networks aided the process of knowledge acquisition to some extent. Evolutionary algorithms were utilised to handle difficult practical engineering optimisation challenges such as production scheduling.

Deep learning (DL), which emerged from neural networks and is based on unstructured data, has advanced in the 2010s, overcoming the symbolic AI bottleneck problem, which is based on knowledge extraction, which is possibly the

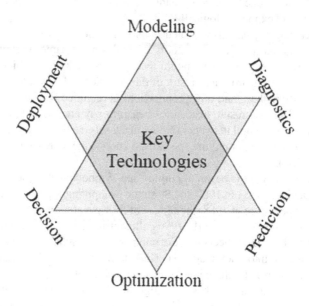

Figure 3.1 Key challenges of artificial intelligence.

most challenging component of constructing an expert system, to some extent. Due to the diversity of AI approaches and solutions, there is no visible distinction between which AI approaches are used. Information is kept in symbolic feature styles in fuzzy systems, but it is implemented as a neuron-like numerical method in neural networks. Intelligent beings can employ both symbolic and subsymbolic methodologies.

3.3 NEURAL NETWORKS

Kohonen defined neural networks as "massively parallel interconnected networks of simple (typically flexible) pieces and their hierarchical organisations designed to interact with real-world items in the same way that biological nerve systems do." "Neural networks are physical cellular systems capable of acquiring, storing, and utilising experimental information," according to Zurada. Neural networks require a thick mesh of compute nodes and connections to attain high performance. Connection models, parallel distributed processing models, and neurophoic systems are some of the other names for them. Figure 3.2 depicts a neural network architecture.

A neural network's essential components are neurons and weighted connections. In a neural network, there are numerous types of model neurons that can be employed. Feldman and Ballard looked into a number of different model neurons.

In general, there are three types of neurons: (1) the p-unit neuron, which is used in the Perceptron and Adline (Widrow and Lehr) models, (2) the q-unit neuron, which is located at the intersection of digital logic and neural networks, and (3) the p-and-q unit, which responds like a p-unit in the normal state but ignores inputs in the recovery state.

A neuron adds up all of its weighted inputs before processing them using a non-linear activation function (nonlinearity). Hard limiters, threshold logic elements, and sigmoidal nonlinearities are the three categories of nonlinearities (Lippmann). See Lippmann and Zeidenberg for more information on model neurons and activation principles (1989, 1990). In multiple directions, signals 8ow can be utilised to recognise neural networks. Neural networks are divided into two types: Deep backward networks and feedback networks. Signals move from an input stage to an output stage via intermediary neurons in a feed-forward network in just one direction.

Signals can pass from every neuron's output to any neuron's input in a feedback network. A property of neural networks is the quantity of instruction received by the learning process from an external actor. Without being taught anything, an unsupervised learning network learns to classify input into sets. Given an input pattern, a supervised learning network adjusts weights based on the difference between output unit values and the teacher's expected values (Zeidenberg).

Neural network models can be classified into six types based on the type of input data (binary or continuous-value) and training methods (supervised or unsupervised) (Lippmann). Figure 3.2 illustrates the taxonomy. For more information, see *Neural Network Models and Learning Algorithms* (Lippmann,

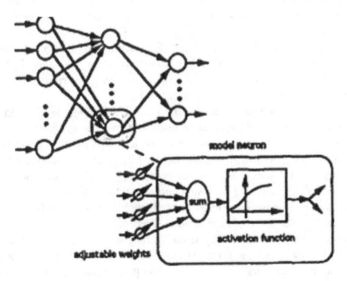

Figure 3.2 A neural network architecture.

Zurada). Traditional computers are not designed to work in the same way as neural networks. As we all know, traditional computers handle data in a serial manner. A serial computer has a single CPU core. The processor receives an instruction and the data it need, executes it, and saves the result to the given memory address.

Everything happens in a deterministic series of operations in such a system. A neural network, on the other hand, is neither sequential nor predictable. A neural network responds to inputs in real time rather than following a set of instructions. The result is the whole state of the network when it has reached some equilibrium condition, rather than a discrete memory location (Caudill). Despite the fact that true neural networks are believed to be quicker and eventually less expensive than conventional hardware (Roberts), neural hardware is still prohibitively expensive. As a result, neural networks have grown in popularity. Ordinary computers are typically used to replicate the process.

3.4 DEEP LEARNING

Smart facilities focus on providing manufacturing intelligence that can have a positive influence across the entire business as new technologies (e.g., IoT, Big Data) are merged into smart manufacturing. Sensory input in a variety of formats, meanings, and structures is bombarding the industry today. Sensory data was gathered from a variety of sources throughout the manufacturing process, including the product line, manufacturing equipment, manufacturing process, labour activities, and environmental variables.

Data modelling and analysis are critical components of smart manufacturing for handling increased high-volume data and enabling real-time operations. Real-time data processing [10,11]. Deep learning has aroused a lot of interest as a breakthrough in computer intelligence, from sensory data to manufacturing intelligence.

Deep learning algorithms are critical for learning from data automatically, finding patterns, and making decisions based on aggregated data. Descriptive analytics, diagnostic analytics, predictive analytics, and prescriptive analytics are some of the types of data analytics that can be done.

Descriptive analytics is to capture the product's settings, surroundings, and operational data in order to summarise what happens. Diagnostic analytics investigate and report on the root cause of a product's performance degradation or equipment failure. Statistical models are used in predictive analytics to create predictions about future output or equipment degradation based on historical data. Prescriptive analytics goes a step farther and suggests one or more actions. Measures can be discovered to improve production results or rectify defects, with the expected consequence of each option indicated.

Advanced analytics enabled by deep learning transform manufacturing into highly optimised smart facilities. Minimising operating expenses, staying on top of changing consumer demand, increasing production while reducing downtime, and improving visibility are just a few of the advantages [12,13]. For global competitiveness, visibility, and getting more value from operations are essential. Deep learning architectures have been established to date, and research subjects in this field are constantly evolving.

Several common deep learning architectures, such as the convolutional neural network, restricted Boltzmann machine, auto encoder, and recurrent neural network, as well as its variations, are researched to aid in the discovery of industrial intelligence. Because these models are the building blocks for developing comprehensive and advanced deep learning techniques, the feature learning capabilities and model creation process were emphasised.

3.5 INDUSTRY 4.0

The rapid development of technology linked to ICT and the internet of things enables manufacturing to expand, resulting in Industry 4.0 [14]. The use of CPS in conjunction with IoT can result in intelligent, flexible systems that can self-learn, which is at the heart of Industry 4.0 [1]. Big data is essential in order to build intelligent and flexible systems. Machine learning, along with data mining, statistics, pattern recognition, and other technologies, plays an essential role in knowledge discovery in databases (KDD) of huge data [15,16]. As part of Industry 4.0's intelligent systems, machine learning is widely used in a variety of industries. The strategies used in manufacturing are meant to derive information from existing data [17]. The new information (knowledge) aids the production system's decision-making and prediction processes. However, the ultimate purpose of machine learning approaches is to bargain decorations in data sets or regularities that describe the links and structure between them [4].

3.6 CONVOLUTIONAL NEURAL NETWORKS

The CNN (convolutional neural network) is a feedforward neural network that is widely used for image and video recognition [18]. The structure of a typical CNN model is shown in Figure 3.3. The input layer builds a feature graph that

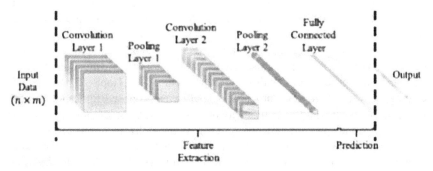

Figure 3.3 Structure of a typical CNN.

corresponds to the convolution kernel using the input vector. The convolution kernel creates a feature graph that is passed on to the next layer using a set of weights. A receptive field, which is a square matrix of weights with sizes smaller than the input, establishes the link between the input and the convolution layer. As it strides or "convolves" along the input region, the receptive field executes the convolution operation.

x(r+1S)(c+jS), where wrc and b signify the weight positioned on the receptive field and the bias, respectively, is the input data element with coordinates (r + I S, c + j S). The nonlinear activation function used to extract features from the input is represented by. Within the convolutional layer, the input size (H WD) is reduced to [(HF+2P S+1) (WF+2P S+1) K], where K is the number of filters. This method continuously reduces the dimension as the convolution layer stack goes deeper. The pooling layer accomplishes two goals: it decreases the spatial dimension of the input layer by up to 75 percent (in most cases) and it prevents overfitting.

3.7 THE EVOLUTION OF DATA-DRIVEN ARTIFICIAL INTELLIGENCE

Artificial intelligence is regarded as a crucial method of intelligence collection, and it was ranked #1 in Gartner's 2017 Top 10 major technology trends [13]. Artificial intelligence has gone through numerous phases since its start in the 1940s to the first and second boom eras in the 1960s and 1980s, ending in the current third boom period (after 2000s). Table 3.1 shows the development trend and common artificial intelligence models. In the 1940s, the MP model and Hebb rule were introduced as the first artificial neural networks to explain how neurons in the human brain worked.

Significant artificial intelligence skills, such as playing chess and solving elementary logic issues, were established at Dartmouth College seminars. The revolutionary work paved the way for the first wave of artificial intelligence (1960s). Perceptron was developed in 1956 as a mathematical model for simulating the neurological system of the human learning system using linear optimisation. The adaptive linear unit network model was developed after that, and it was effectively employed in real-world applications such as communication and weather forecasting in 1959.

Early artificial intelligence's inability to perform nonlinear issues like XOR (or XNOR) categorisation was also noted as a shortcoming. Artificial intelligence advanced to the next level with the development of the Hopfield network circuit. (1980s). In 1974, back propagation (BP) was proposed for handling nonlinear problems in complex neural networks. In 1985, a random mechanism was introduced to the Hopfield network, resulting in the Boltzmann machine (BM).

A support vector machine (SVM) was constructed with kernel functions modification when statistical learning was introduced in 1997, and it performed well on classification and regression. Traditional machine learning algorithms require human expertise for feature extraction in order to limit the dimension of input, hence their efficiency is heavily dependant on specified features. Deep learning is based on a variety of classic machine learning methodologies, as well as statistical learning as a source of inspiration. To complete tasks, deep learning uses data representation learning rather than explicit constructed characteristic learning.

It converts data into abstract representations that can be used to learn features. In 1986, the restricted Boltzmann machine (RBM) was created by getting the Boltzmann machine's probability distribution and using the hidden layers as feature vectors to characterise the input data. Meanwhile, auto encoder (AE) was developed to reduce the loss function using the layer-by-layer greedy learning method. The recurrent neural network (RNN) was established in 1995 as a neural network for feature learning from sequence data with directed topological connections between neurons.

Long short-term memory (LSTM) was created in 1997 as an upgraded type of RNNs to deal with complex time sequence data and solve the vanishing gradient problem. In 1998, CNNs were proposed as a way to handle two-dimensional inputs (such as photographs), with features learned by stacking convolutional layers and pooling layers. Model training and parameter optimisation grow increasingly difficult and time expensive as the hierarchical structures of deep learning models become more intricate, perhaps leading to overfitting or local optimisation difficulties.

The sparse auto encoder (SAE) was developed a year later to reduce dimensionality and learn sparse representations. Deep learning is rapidly gaining popularity. In 2009, a bidirectional deep Boltzmann machine was built to learn ambiguous input data consistently, and the model parameters were improved via layer-wise pre-training. Denoising auto encoder was initially released in 2010, with the purpose of reassembling stochastically distorted input data and forcing the hidden layer to locate more resilient features.

The deep convolutional neural network (DCNN) with a deep convolutional neural network structure was released in 2012, and its image recognition performance was improved.

In 2014, the generative adversarial network (GAN) was proposed, which consists of two adversarial models. The discriminative model was used to train and classify both actual and generated random samples, while the generative model was used to generate random samples that were comparable to real examples. By combining attention mechanisms with LSTM, an attention-based LSTM model was suggested in 2016. In today's world, new models are being developed on a weekly basis.

3.8 LSTM NEURAL NETWORKS

The LSTM is a type of RNN that keeps the temporal dimension of sequential data by linking neurons to form a network that is a direct cycle of the input data. The long-term state component c(t) and the short-term state component h are

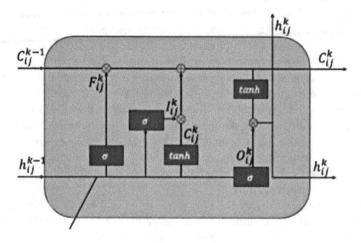

Figure 3.4 Basic architecture of LSTM memory card.

shown as two separate components of an LSTM memory cell in Figure 3.4. (t). As shown in Figure 3.4, the memory cell has three control gates: input, output, and forget gates, which perform write, read, and reset operations in each cell, respectively.

The multiplicative gates allow the model to store information over long periods of time, avoiding the problem of vanishing gradients that plagues RNNs [12]. The input, forget, and cell activation, as well as the output vector, are defined by the following equations, which allow the LSTM to predict the output vector: The input, forget, and cell activation, as well as the output vector, are defined by the following equations, which allow the LSTM to predict the output vector:

$$it = \sigma \ (Wxixt + Whiht -1 + Wcict - 1 + bi) \tag{3.4}$$

$$ft = \sigma \ (Wxfxt + Whfht -1 + Wcf \ ct - 1 + bf) \tag{3.5}$$

$$ct = ftct - 1 + itg \ (Wxcxt + Whcht - 1 + bc) \tag{3.6}$$

$$ot = \sigma \ (Wxoxt + Whoht - 1 + Wcoct + bo) \tag{3.7}$$

$$ht = oth \ (ct) \tag{3.8}$$

W and b stand for weight matrix and bias vector, respectively, while (.) stands for a conventional logistic sigmoid function. The input gate, forget gate, output gate, and cell activation vector are represented by the variables I f, o, and c, respectively (Figure 3.5).

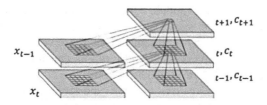

x_{t-1}

x_t

$t+1, c_{t+1}$

t, c_t

$t-1, c_{t-1}$

Figure 3.5 Inner structure of CNN LSTM.

3.9 DATA-DRIVEN SMART MANUFACTURING

Data-driven smart manufacturing has a wide range of ramifications. Big data analytics is used by manufacturing companies to analyse data from their operations in order to improve the flexibility and intelligence of their processes. Manufacturing has changed from primary to smart processes as a result of fully utilising industrial data. This improves the efficiency and productivity of the factory. Big data technology is used to collect, store, process, and evaluate manufacturing data. As a result, manufacturing intelligence will be greatly enhanced.

The production module, the data driver module, the real-time monitor module, and the problem processing module make up the data-driven smart manufacturing framework, as shown in Figure 3.3.

(a) *Manufacturing module*: This module is designed to perform a wide range of manufacturing duties. It is made up of a number of different information systems and industrial resources that can be summed up as man-machine-material-environment. Raw materials are the inputs to this module, and completed goods are the outputs.

Data is acquired from human operators, production equipment, information systems, and industrial networks during the input-output transformation process.

(b) Data driver module: This module is the engine that propels smart manufacturing forward at all phases of the production data lifecycle. Data from the production module is sent to cloud-based datacenters to be processed further. The operations of the manufacturing module are then led by accurate data and actionable recommendations derived from various types of raw data (for example, product design, production planning, and manufacturing execution).

To address specific product quality issues, the manufacturing process can be altered. As a result, the real-time monitoring module can aid in the more efficient operation of industrial facilities.

(c) The problem processing module identifies and forecasts upcoming issues (such as equipment breakdowns or quality issues), diagnoses underlying causes, suggests feasible solutions, estimates remedy efficacy, and assesses potential repercussions on other production operations.

Human operators and artificial intelligence programmes can not only make informed judgements based on real-time data and analysis of previous and ongoing data provided by the data driver module, but they can also prevent similar problems in the future. This module's proactive maintenance will improve the efficiency of industrial processes. Data collection, integration, storage, analysis, visualisation, and application are all standardised processes that may be used to a wide range of sectors. The proposed data-driven smart manufacturing architecture is designed to be extensively useful in this setting.

Depending on resource availability, small and large businesses can employ different techniques to achieve data-driven smart manufacturing at different scales. Small and medium-sized enterprises (SMEs) can use on-demand cloud computing services provided by third parties such as Amazon and Alibaba, in contrast to larger enterprises that can afford to construct their own cloud infrastructure for data storage and analysis. All of the production activities and resource allocation can be closely synchronised to support smart manufacturing.

3.9.1 Characteristics of data-driven smart manufacturing

The five characteristics of data-driven smart manufacturing are as follows:

(1) It facilitates customer-centric product development by incorporating user data into product design. Big data analytics, for example, can be used to precisely evaluate consumer demographics, needs, preferences, and behaviours, allowing for more personalised products and services to be created.

(2) It promotes self-organisation by taking advantage of manufacturing resources and task data to plan more efficient output. Internal and external data from a range of industrial facilities, for example, can be used to build production plans. The best manufacturing resources are picked to develop the optimal configuration that meets all of the manufacturing activity's criteria in order to put production plans into action.

(3) It allows precise management of a range of manufacturing process factors, making self-execution easier. Appropriate raw materials and parts, for example, can be delivered at any time to any production facility that requires them, and manufacturing equipment can mill raw materials or assemble parts as needed.

(4) It allows for self-regulation by monitoring production processes and providing real-time status information. A manufacturing system, for example, can respond automatically to unforeseen events by allowing AI systems as well as human operators to programme its behaviour (e.g., a scarcity of manufacturing resources or a change in manufacturing tasks).

(5) It uses historical and real-time data for proactive maintenance and quality control, allowing for self-learning and adaptation. Machine failures and quality concerns, for example, can be predicted and avoided before they occur, allowing manufacturing processes to adjust to future issues proactively. Finally, data-driven smart manufacturing offers a wide range of services to

manufacturing companies. One of the most visible advantages is the possibility to boost manufacturing efficiency and product quality significantly. The paradigm is best illustrated through concrete applications that take into account the aspects stated above as well as the industrial data lifecycle.

3.9.2 Data-driven smart manufacturing application

Manufacturing is the process of transforming raw materials into finished products and value-added services through the coordination of appropriate manufacturing facilities, resources, and activities. Smartdesign, smart planning, material allocation and tracking, manufacturing process monitoring, quality control, and smart equipment maintenance are just a few of the most exciting applications that may be implemented during the production process.

3.9.3 Smart design

The importance of design cannot be understated because it influences the majority of a product's manufacturing costs. Product design is trending toward data-driven design in the big data era [4]. Researching and understanding client requirements, behaviours, and preferences is the first step in product creation. This type of information can be found both online and offline. Customers are becoming more adept at using portals such as social networking sites, ecommerce platforms, and product/service review sites to share first-hand experiences with products on the Internet [2].

Rich user data (e.g., biological data, behaviour data, and user–product interactions) may now be available from a rising number of IoT-connected smart objects (e.g., smart phones and wearable devices). Manufacturers' capacity to turn customer opinions into product features and quality needs improves when they take a holistic approach to exploiting user-related big data. It also allows designers to streamline design processes, increase product innovation, and create more individualised goods for customers. [3,4,5,6,7,8,9,14,15,16,17,18,19,20].

Furthermore, when compared to traditional approaches (e.g., interviews, surveys, etc.), big data analytics helps customers to save money while speeding up computationally expensive operations (e.g., market preferences and customer demand analysis, etc.) [5].

3.9.4 Intelligent planning and process improvement

Production planning is essential even before a product is developed to determine a manufacturing facility's production capability as well as the availability of resources and suppliers.

Production planning and shop floor scheduling can both benefit from big data analytics. To begin, big data analytics techniques are applied to a variety of data, such as customer orders, manufacturing resource status, production capacities, supply chain data, sales data, and inventory data. Hypernetwork-based manufacturing resource supply-and-demand matching and scheduling [7] can be carried out using the information collected from these technologies to quickly discover available resources. The ideal arrangement of manufacturing resources is then discovered using smart optimisation techniques, which are then used to develop production plans. Procedures for completing the task [8,9].

Furthermore, prior to the start of production, process optimisation is critical. Big data analytics is beneficial for assessing and optimising technical processes. To study the link between distinct technological factors and their effect on yield and quality, several forms of process data can be employed, including historical data and data on the patterns and correlations inherent to certain processing steps. Adapting technological processes to these parameter scans improves product quality while increasing efficiency and lowering costs.

3.9.5 Material distribution and tracking

Material distribution is determined by production planning and progress, as well as any on-site emergency requirements. In an ideal world, the appropriate material would be delivered to the appropriate equipment at the appropriate moment, allowing it to be processed according to the appropriate protocols. A range of material-related data, such as inventory, logistics, and progress data, can be regulated to achieve this purpose. Material flow data from various sources is compared to material data (e.g., data from human operators, machines, vehicles, etc.).

These studies can be used to determine material distribution in terms of material type, quantity, delivery time, and technology, allowing for the most efficient production logistics. To ensure seamless production, material can be delivered on time, according to the real production pace and conditions (i.e., avoiding unnecessary production delays, interruptions, or production stoppages). Material traceability [1] is also necessary to guarantee that different types of materials meet their own set of quality criteria and standards. During the production process, identification tags can be used to track material conditions (such as location, status, and quality) in real time.

In AGV [2], for example, an RFID-enabled positioning system enables effective material distribution throughout industrial areas. Operational data can be created using big data analytics for product quality control and fault tracking throughout the manufacturing process.

3.9.6 Manufacturing process monitoring

A variety of things have a role in the manufacturing process. Certain factors can influence the manufacturing process and product quality discrepancies (such as production equipment, materials, the environment, and technological characteristics). They also have the ability to converse with one another. As a result, real-time monitoring of all phases of the production process is crucial.

However, determining which variables have a systematic impact on manufacturing processes can be difficult. Thankfully, big data can assist with technical monitoring of manufacturing operations.

The best acceptable design range for each production element can be proposed using big data analytics' predictive capability. An issue is found when a factor deviates from its permissible range, and operators receive alerts and advice to make immediate corrections, ensuring higher production process uniformity. Anomalies (for example, order lateness) are frequently produced by unusual events such as equipment failure, material shortages, and operation variation, to mention a few.

Anomalies in manufacturing frequently reveal patterns that can be detected using a variety of time series data (e.g., material consumption data, energy consumption data, rotation rate, vibration, torque, etc.). Because such input is frequently time-dependent, static models are unable to process it [3]. Furthermore, traditional data analysis approaches are computationally intractable and incapable of processing large amounts of data [3]. By combining the aspects of time and causation [4], an early-warning model of production anomalies on the shop floor can be created using appropriate big data approaches such as decision tree (e.g., ID3 and C4.5) and neural network [5,6].

It is feasible to forecast whether and when production abnormalities will occur in advance by mining the feature patterns and trend of abnormal occurrences in time series. When working with multi-source data and large data sets, big data analytics offers for greater flexibility, accuracy, and computation time.

Manufacturing processes can be dynamically altered based on big data analytics while maintaining a balanced use.

3.9.7 Control of product quality

For smart manufacturing, many data-driven quality control solutions are being developed [7]. Sensors, RFIDs, and machine vision applications can be used to collect product quality data such as geometric parameters (e.g., thickness, length, and surface rheology).

As a result, product quality faults can be identified, diagnosed, and addressed very instantly. Data mining and data integration can be used to uncover less visible sources of production inefficiencies, such as poor equipment connections and procedures. As a result, not only may low-quality or failed products be discovered and removed automatically, but quality-related variables can also be eliminated or managed. Big data analytics, when combined with machine learning, will also allow manufacturing companies to undertake case-based reasoning. Lessons learnt in one quality control scenario can be applied to a variety of situations.

3.9.8 Smart equipment maintenance

Data analytics may be utilised to properly estimate and diagnose equipment faults and component lifetimes, resulting in better maintenance decisions. Big data analytics can anticipate the tendency for equipment capacity to diminish, component longevity, and the source and degree of individual failures by combining smart sensor data with domain knowledge, past experience, and historical records about equipment maintenance [6]. Equipment failure trends can be detected using big data analytics in a variety of ways, including seasonal, periodic, combinational, and other patterns.

This information can be utilised to avoid errors by taking preventative steps. As a result of big data analytics' predictive capabilities, the equipment maintenance paradigm has shifted from passive to proactive, extending equipment life and lowering maintenance costs [17]. Energy consumption [18] can also indicate equipment malfunctions or anomalies. Big data combined with energy consumption can help spot energy changes, irregularities, or peaks in real time by constructing a multidimensional energy consumption analysis model. The surrounding industrial processes, equipment, and energy supply can all be dynamically regulated to assure regular production.

3.9.9 Dataset framework and detection

The amount of dynamically changing data created during a product's lifecycle is increasing. The information gathered can be used to improve the manufacturing company's efficiency. In three ways, this chapter has aided smart manufacturing.

(1) A historical perspective: Manufacturing data has evolved over four epochs of production: handicraft, machine, information, and big data.
(2) From a development aspect, the enormous industrial data lifecycle was described as a succession of phases, including data production, collection, transmission, storage, and integration, processing and analysis, visualisation, and application.
(3) A look at the future of data in the manufacturing industry.

There are a variety of constraints to consider. To begin with, today's data gathering systems aren't entirely prepared for smart data perception, especially when working with heterogeneous devices that communicate via diverse interfaces and protocols [16]. Second, while cloud-based data storage and analytics has been demonstrated to be a viable technological solution, its applicability for low-latency and real-time applications is limited due to unresolved issues (e.g., network unavailability, overburdened bandwidth, and unacceptable latency time, among others) [20].

Third, while it is widely recognised that the integration of the physical and cyber worlds is an important characteristic of smart manufacturing, the vast majority of past research has concentrated on data acquired from physical models rather than data from virtual models [17]. On the other hand, this research offers a first glimpse into data-driven smart manufacturing and its prospective uses. There are a few intriguing study avenues that interested scholars can follow in the future.

(1) The data-driven smart manufacturing framework may contain critical technologies for data perception and collection from heterogeneous equipment, such as IoT gateways or industrial Internet hubs.

Data collection and transfer will be easier with devices that are interoperable with a variety of interfaces and communication protocols.

(2) The suggested architecture is adaptable enough to accommodate emerging data storage and processing techniques such as fog computing and edge computing.

Fog computing and edge computing can bring a manufacturer's data compute, storage, and networking capabilities closer to the edge, reducing bandwidth requirements, latency times, and service downtime significantly [6].

(3) Digital twin technologies can be accommodated by the proposed framework. A digital twin can be used by manufacturers to handle real-time, bidirectional, and coevolving mapping between a physical entity and its digital representation, paving the path for deep cyber-physical integration. When data-driven smart manufacturing is combined with a digital twin, the result is more adaptable, predictable, and flexible manufacturing.

References

1. G. Revathy et al., Revelation of Diabetics by Inadequate Balanced SVM, *Turkish Journal of Computer and Mathematics Education*, Vol 12, no 2, 2021.
2. G. Revathy et al., Firefly Optimization in IOT Applications for Wireless Mesh Networks, *Turkish Journal of computer and Mathematics Education*, Vol 12, no 2, 2021.
3. G. Revathy et al., Brain Tumor Detection in MRI Images, using Fuzzy C-Means and Cuckoo Search Algorithm, *Solid State Technology*, Vol 63, Issue 5, 2020(Anex-1)
4. G. Revathy et al., Indian Traffic Road Sign Recognition for Intelligence Driver Assistance System Using SVM, *Journal of Critical Reviews*, Vol 7, issue 9, 2020.
5. G. Revathy et al. Bi-Directional Wind Turbine for Power Generation in a DC Micro Grid, *Journal of Critical Reviews*, Vol 7, issue 6, 2020.
6. G. Revathy et al., Enhancing Security and Efficient Authentication Scheme using K means Clustering, *International Journal of Engineering Research and Technology*, Special issue 2020.
7. G. Revathy et al., Implication of IoT With Security in Wireless Mesh Networks, *International Journal of Future Generation Communication and Networking*, Vol 13. Issue 2, June 2020.
8. G. Revathy et al., Human Fingerprint Recognition System (HFRS) For Real-Time Application Using Support Vector Machine (SVM), *International Journal of Advanced Science and Technology*, Vol 29, Issue 6, May 2020.
9. G. Revathy, Anju, Parimalam and Imaya Kanishka, Diabetic Detection Using Irish, *International Journal of Scientific Research in Engineering and Management (IJSREM)*, Vol 4, Issue 3, March 2020, ISSN: 2582–3930.
10. S.T. Ahmed, S. Sreedhar Kumar, B. Anusha, P. Bhumika, M. Gunashree and B. Ishwarya (2018, November). A Generalized Study on Data Mining and Clustering Algorithms. In International Conference on Computational Vision and Bio Inspired Computing (pp. 1121–1129). Springer, Cham.

11. S.T. Ahmed, D.K. Singh, S.M. Basha, E.A. Nasr, A.K. Kamrani and M.K. Aboudaif (2021). Neural Network Based Mental Depression Identification and Sentiments Classification Technique from Speech Signals: A COVID-19 Focused Pandemic Study. *Frontiers in Public Health*, 9.

12. K. Periasamy, S. Periasamy, S. Velayutham, Z. Zhang, S.T. Ahmed and A. Jayapalan (2021). *A Proactive Model to Predict Osteoporosis: An Artificial Immune System Approach*. Expert Systems.

13. V. Sathiyamoorthi, A.K. Ilavarasi, K. Murugeswari, S.T. Ahmed, B.A. Devi and M. Kalipindi (2021). A Deep Convolutional Neural Network Based Computer Aided Diagnosis System for the Prediction of Alzheimer's Disease in MRI Images. *Measurement*, 171, 108838.

14. G. Revathy, N.S. Kavitha, K. Senthivadivu, D. Sathya and P. Logeshwari, Girl Child Safety using IoT Sensors and Tabu Search Optimization, *International Journal of Recent Technology and Engineering (IJRTE)*, Vol 8, January 2020, ISSN: 2277–3878.

15. G. Revathy, G. Saravanan, R. Madonna Arieth and M. Vengateshwaran, Magnify QoS with Tabu and Link Scheduling in WMN, *International Journal of Recent Technology and Engineering (IJRTE)*, Vol 8, Issue 4, November 2019, ISSN: 2277-3878.

16. G. Revathy, Mounting Eminence of Services in Wireless Mesh Networks, *International Journal of Research and Analytical Reviews*, September 2018, ISSN 2349 5138.

17. G. Revathy and K. Selvakumar, Sustain Route by Tabu and Amplified QoS by Distributed Scheduling in WMN, *International Journal of Recent Trends in Enginnering and Research*, ISSN: 0973-7391.

18. G. Revathy and K. Selvakumar, Channel Assignment Using Tabu Search in Wireless Mesh Networks, *Wireless Personal Communication*, ISSN 09296212.

19. G. Revathy and K. Selvakumar. Increasing Quality of Services in Wireless Mesh Networks, *International Journal of Advanced Research in Computer Engineering and Technology*, Vol 7, Issue 3, March 2018. ISSN 22781323.

20. G. Revathy and K. Selvakumar, Escalating Quality of Services with Channel Assignment and Traffic Scheduling in Wireless Mesh Networks, *Cluster Computing*, Jan 2018. ISSN 13867857.

Chapter 4

Heterogeneous data management in IoT-based health care systems

Geetanjali Kale and Kalyani Waghmare

CONTENTS

4.1 INTRODUCTION

Health is an important concern for everyone, right from babies in their mother's womb to old-age people. There is an intense need for regular health check-ups and analysis of health reports to investigate the health status of individuals.

It demands massive availability of medical experts, specialist doctors and services. According to an article published in *India Today* on 29 April 2016, India has 1:2,000 doctor-to-patient ratio and the World Health Organization (WHO) recommended that it should be 1:1,000 as reported in Mezghani et al. (2015). In rural areas, the availability of doctors is generally rare and very few expert doctors are available. Nowadays people are very keen and conscious about their health. Due to the availability of various electronic gadgets like health bands, handy blood

DOI: 10.1201/9781003335801-4

pressure, sugar checkup and pulse recording devices, people are doing analysis of their own and trying to find solutions or prefer taking precautions at the early stages of diseases. This demands automated smart medical assistance systems for the medical diagnosis of patients, which could be accessible from the Web or mobile at anytime from anywhere.

The health care systems getting information from various sources such as pathological report, electrocardiogram (ECG) and electroencephalogram (EEG) have different processing requirements than computed tomography (CT) scan, ultrasound, magnetic resonance imaging (MRI), functional magnetic resonance imaging (fMRI), mammography, sonography or X-ray. The history of the patients is maintained by doctors in the form of structure. Ample amount of such heterogeneous data is generated frequently.

As data are generated from various sources, distributed systems are preferred with security and consistency. Currently, many health care systems are developed using blockchain as sharing of information in the blockchain happens in a secure and transparent way. In the health care system, different use cases are available like patient electronic health record management, online patient–doctor interactions and claims management. The platform developed by BURSTIQ is securely managing health and identity data. It is developed on a blockchain platform by following Health Insurance Portability and Accountability Act (HIPAA) rules, which are explained in Edemekong et al. (2018) and Nosowsky et al. (2006). Factom software from Austin developed a blockchain-based product to securely store digital health records. Medicalchain is a platform which gives patients a secure way for storing medical data and health care professionals get access to personal health data. The platform makes interaction possible.

Is health care system a necessity or just a facility? Maybe before COVID-19, in a few countries having a huge population, it could be considered as just a facility but during COVID-19, all the nations realised the importance of health care systems.

Health care systems are not just about storing and maintaining the data, but it comprises the maintenance and improvement of medical services to supply the medical aid on demand for people efficiently and effectively. The fast-growing demand of medical aid by patients emerged from advancements made in health care systems. Countries strive for betterment of health care systems and are undertaking rigorous transformations.

Health care systems are multidisciplinary systems involving domains like Internet of Things, computer vision, text analysis, machine learning, artificial intelligence, cloud computing, Web technology, mobile technology and many more. As discussed earlier, a huge amount of data is generated through various sources and sensors. This data needs to be saved securely by preserving the privacy of the personal details.

We discuss the architecture of typical health care system in Section 4.2. Section 4.3 provides the details of data preprocessing and storage. Further, the algorithms for intelligent decisions are discussed in Section 4.4. The challenges and scope for

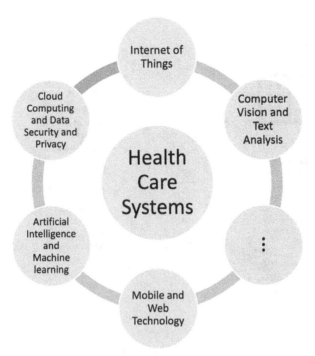

Figure 4.1 Various domains in multidisciplinary health care systems.

further research in health care domain are discussed in Section 4.5. The chapter is concluded with a discussion and conclusion.

4.2 ARCHITECTURE OF TYPICAL HEALTH CARE SYSTEM

Figure 4.2 shows the architecture diagram for a typical health care system. Generally in these kinds of systems, users need to be always connected with the wearable devices and further with data servers through the internet or intranet. Wearable devices are bands that can sense and record even small activities, pulse rate, blood pressure and temperature tracking and these can be worn on the wrist, chest or back. Also the data is collected through various other IoT devices. All the users of the system including doctors, medical experts, exercise trainers and patients should have their personal account in the system possibly connected with a unique identification number to preserve personalisation and privacy. Every user has a personalised interface and functionalities available in their login. Experts, doctors and trainers can monitor all their patients' progress, the schedule adherence and also own health details. Patients' details and medical history of patients can be maintained on a Hadoop cluster, personal cloud or public cloud environment.

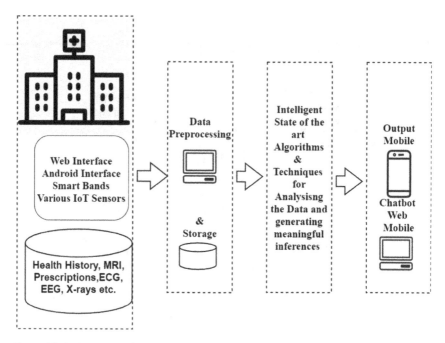

Figure 4.2 Architecture of typical smart health care system.

Both Android and Web interfaces are created for convenient interaction as per the availability of devices. All the body parameters collected from various sensors, history information and pathological reports should be maintained at the central place. Generally, doctors recommend preliminary examinations like blood pressure, temperature and heart beats and ask some questions for diagnosis of the patient's condition. These parameters can be collected through various sensors and interactive interfaces. Smart systems can be designed to predict probable diseases and vision-based monitoring and voice interaction can be designed for the patients that require 24×7 attention. Voice systems can also be used for interactive entertainment of patients.

Hadoop cluster will maintain all the data collected in the form of numbers, text, images and videos. This data will be analysed and results like the current state of health and probable diseases can be obtained.

Personalised plans for every user of the system will be prepared and the major focus of the system will be on preventive remedies and exercises. Vision-based exercise monitoring system is capable of assessment of exercise will be designed and it will keep track of schedule as well.

These kinds of automated smart health care systems will be beneficial for various government agencies to gather different health related statistics to implement their schemes. The system will be designed by keeping various security aspects in mind and personalised data security will be maintained using block chain technology. Intelligent state-of-the-art machine learning and data analysis algorithms will help to draw meaningful inferences from this data. These results will be further used or communicated to the system users through various interfaces like the Web and Android.

4.3 DATA PREPROCESSING AND STORAGE

Data preprocessing is the main step in machine learning applications. In this phase, data is prepared for mining. Duggal et al. (2016) demonstrated that the use of proper data preprocessing methods improves the results of health care decision systems. The main steps followed in data preprocessing are data cleaning, data integration, data scaling with normalisation and feature reduction.

4.3.1 Data cleaning

The data in the health care system is collected from various sources, so it is very rarely clean and complete. This may be due to network connectivity or the device from which readings are taken may be erroneous, or human errors while feeding

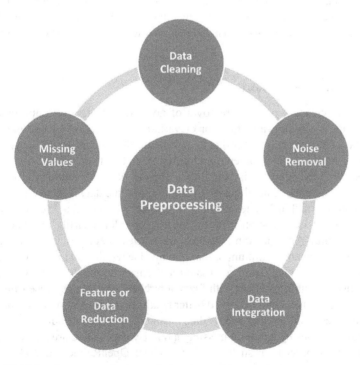

Figure 4.3 Major data preprocessing techniques.

data or issues while maintaining atomicity, consistency, integrity and durability of data. In the data cleaning step, all these missing values, errors and inconsistent issues are usually handled. Ayilaraet al. (2019) observed that proper handling of missing data extensively increased the precision of estimated change in patient-reported outcome scores.

4.3.2 Missing values

Consider in our database we have "WBC-count" as one of the attributes. There are some tuples where WBC-count is missing. We may ignore these tuples while training our model. But ignoring is the worst solution and results in a hamper classification. These missing tuples could be replaced manually. But manual replacement is a very time-consuming and complex solution for large data sets. The other way could be to replace it by using statistical measures like mean, median, mode or central tendency. The preferred method is calculating feasible value by applying regression, inference-based algorithms like decision tree, Bayes theorem and maximum likelihood method for imputing the value. While handling missing value, the main challenge is that we need to study carefully central tendency and dispersion of data. The choice of the measure must be done considering skewness of data, outliers and distribution. The missing attribute is a class label. For example, we have a skin disease data set considering some eight diseases and in our dataset for few tuples, the class name is not mentioned. In this scenario, we can use clustering technique to label the class name for the missing tuples.

4.3.3 Noise removal

In the data cleaning process, removal of noisy data is also an important task. Noisy data is actually random error or variation that occurred while measuring the data. This error may occur due to calibration of different instruments such as weighing machine, thermometer, and pulse oximeter. The well-known and easy method for noise removal is "binning". Binning performs smoothing of sorted data by considering its neighbours. In binning, first the data is sorted and divided into a number of "bins". The decision of bin size is challenging. So to decide the bins, graphical visualisation methods such as boxplot, quartile plot and scatter plot are helpful. After deciding the bins, the entries in every bin will be replaced by mean, median and minimum or maximum value within that bin. Noise removal can also be performed using regression and clustering methods as well. Hasan (2013) used a binning method with fuzzy inference for reducing noise from ultrasound images. The author observed better results than the existing methods. So in data preprocessing, the first step is data cleaning, which contributes to removing data discrepancies by handling missing and noisy data. There are many commercial and open-source tools such as Potter's Wheel, OpenRefine and Drake that are available for removing noise in the data set.

4.3.4 Data integration

In health care systems, data comes from heterogeneous platforms. So semantic and structural heterogeneity is an excessive challenge in the data preprocessing phase. In health care systems not only inputs are coming from various sources but various stakeholders such as patients, doctors, health care workers and insurance agents are involved during various functions. Entity identification and redundancy are the major issues faced during data integration in health care systems. Consider we have reports coming from pathology having Patien_id as a field and P_id as a field mentioned in the radiologist report. The names of the fields are different but storing the same information. So while integrating data from different sources, users have to study the metadata, which includes other information about the entity like meaning of data, datatype, range and so on. The metadata helps in identifying entities as well as for reducing redundancy and transformation of data. Redundancy is a major issue in integration and it is sorted by performing a correlation analysis. To check redundancy in numerical features like weight, height and temperature, covariance or Pearson correlation coefficient is used. Higher the value of coefficient stronger is the correlation. If height and weight are strongly correlated, then one of them may be removed from that data. The correlation among nominal features like gender, designation and grades could be calculated using chi-square method. In this era of technology, many platform-dependent and independent APIs are developed and evolved for making efficient and effective data integration. Cong Peng and et al. (2020) performed an extensive survey on health care data integration with respect to current trends in technologies. There are many open-source tools like Apache Kafka, Pentaho Kettle and Talend Open Studio readily available for data integration. Mainly data warehouses or data marts are implemented for data integration in health care systems.

4.3.5 Feature or data reduction

The availability of Web Interface, Android Interface, electronic gadgets, distributed systems and IoT sensors made it easy to collect data from various sources. But the consideration of the number of features is not necessarily directly proportional to the performance of classification or prediction. Jain and Singh (2018) performed dimensionality reduction to improve the performance of classification and clustering algorithms. Feature reduction includes two main methods: dimensionality reduction and numerosity reduction. Principal component analysis (PCA) and attribute subset selection are well-known methods effectively used for dimensionality reduction in many health care systems such as the health system capacity assessment system in rural Kenya developed by Wanzala et al. (2019) and heart disease predicting system developed by Spencer (2020). Panay et al. (2019) performed the numerosity reduction to convert voluminous data into smaller form by applying parametric methods such as regression. Salem et al. (2019) constructed histograms with equal width and equal depth or

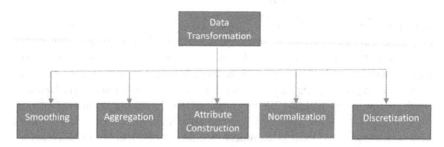

Figure 4.4 Data transformation methods.

non-parametric methods such as clustering and variant sampling ways. Optimal feature selection or feature reduction is an important task in the data preprocessing phase of machine learning for improving the results.

4.3.6 Data transformation and discretisation

In health care system, data is heterogeneous as well as unstructured. Data transformation is done to make this data homogeneous and structured. Figure 4.3 shows the various methods used for data transformation.

Data smoothing methods remove noise from data using binning, regression and clustering. In data aggregation, the data from various sources is collected, analysed and presented in a summarised form. The use of a data cube from a data warehouse is recommended while doing aggregation. Many times while transforming the data, new attributes are constructed by combining more than one attribute in the dataset. For example, instead of keeping separate attributes for location, city, state, country and pincode, only a single attribute "Address" is created. Normalisation converts all the numerical values of any attribute in the given range. The effect of normalisation could clearly be seen with machine learning techniques. Pathak et al. (2013) discussed use of normalisation in electronic health records for high throughput.

4.3.7 Data discretisation

Health care system takes many inputs that are continuous or in time-series form. Many IoT devices such as smart watches, thermometers, health bands and many more create continuous streams of data. The EEG, ECG, MRI and CT scan results generate time-series data. The discretisation of continuous and time-series data is necessary to increase the error-free data. Also it may happen that continuous values may not correlate with output variables. So binning is the well-known and basic method of discretising data effectively. Discretising in small bins might correlate with output variables. If we plan to work with machine learning models then, there are a few models such as decision trees and neural networks in which continuous attributes are not acceptable. Acosta-Mesa et al. (2014) performed discretisation of time-series data using evolutionary algorithms for the classification of precancerous cervical lesions.

4.3.8 Data storage

In the health care system, data is heterogeneous. So its storage and retrieval is a very challenging task. Usually data warehouse and service-oriented architecture (SOA) frameworks are preferred for efficient data retrieval. To build a data warehouse, business and technological skills are necessary. Business skills are required to understand which entities like administration, finance, radiology sources, pathology sources, insurance claims and so on need to be considered and also what type of queries are expected to be solved. The technological skills are required to understand how to make assessments from quantitative data and derive knowledge from historic information in the data warehouse. To build enterprise warehouse models for health care from scratch, you need to think carefully at the start for all the scenarios while designing the database. The size of this database may be then very huge in terabytes or petabytes. The major disadvantage of building an enterprise warehouse is it may be too expensive because of the high-end server requirement and skilled human resources. Also accommodating changes in design will be very difficult. Instead of enterprise warehouse, data marts are preferred. Data mart for every subject like administration, radiology department and finance department will act independently as well as integration of it is also possible easily. The OLAP cubes of data warehouse effectively handles the queries on historical data while the distributed file systems effectively work on queries of online transaction processing (OLTP) functions.

In recent years, many applications like fake news detection, supply chain, logistics monitoring and voting systems are successfully implemented in blockchain. Monrat et al. (2019) did a survey of applications, challenges and opportunities. As an effect of this success, researchers introduced the use of blockchain in the field of medical health care to solve security issues such as integrity and data authenticity. The use of blockchain technology keeps data secure and free from spying by maintaining distributed ledger to record transactions between parties in decentralised structure. The main advantage of blockchain is that it provides continuous service because of its decentralised structure. Another advantage of blockchain is that it provides security by using hashing and the proof-of-work (PoW) and proof-of-stake, which are hard to tamper with transactions stored in the blocks. Wang (2019) reviewed consensus mechanism and mining strategy in blockchain network management.

4.3.9 Data imbalance in health care

Data imbalance is an important issue that arises in health care data for some rare kind of diseases, symptoms or improper selection of population. This may lead to improper training of your machine learning model showing great accuracy but giving wrong classification. Some of the solutions to handle data imbalance are:

- Over-sampling (up sampling)
- Under-sampling (down sampling)
- Feature selection
- Use the right evaluation metrics like confusion matrix, precision, recall and F1-score.

4.4 BIG DATA PROCESSING IN HEALTH CARE

The vast amount of heterogeneous medical data generated in today's digital world needs to be processed and managed to provide better health delivery. Health is an important concern for everyone and health care is a very complicated inter-disciplinary domain. It involves issues right from psychological disorders to road accidents. Patients' history plays an important role in any treatment, so maintaining a health record is a very important aspect of a smart health care system. These records are heterogeneous in nature and are maintained in the form of videos, audios, text and images of X-rays, EEG, ECG, and so on and text data in the form of summary reports and prescriptions. This gives rise to big data and it has many challenges related to storage, personalisation, security, analysis, trans-formation and integration.

There are many systems that make use of the state-of-the-art big data analytics techniques for predictive analysis of diseases like heart attack and cancer. Big data analytics in medical imaging helps in the analysis of various kinds of images for early understanding of the disease. Kanitkar et al. (2019a, b, c) have applied various machine learning techniques on the cervical images and discussed their comparative results. These kinds of advancements in the medical domain will upsurge the availability of the intelligent systems at rural places as well and it will reduce the burden on highly skilled medical staff.

In the COVID-19 pandemic situation, large amounts of data are being generated worldwide through various sources. This heterogenous data requires large amount of multidimensional data to be stored securely to maintain privacy and it also requires the high-performance computing for processing and analysing the CT images that provides the severity of the lung infection due to COVID-19. This kind of analysis helps in the early detection of the infection and avoids further complication.

Advances in big data analysis, storage, communication and visualisation will definitely help the medical industry to generate more smart systems not only for curing the diseases but for preventing many diseases and complications.

4.5 ALGORITHMS FOR INTELLIGENT DECISIONS

Researchers are working on computer-assisted health care systems worldwide. Due to lack of availability of medical experts and increasing cases of medical emergency, there is high demand for such systems. IBM initiated the design of a health care assistive system for doctors using IBM Watson API. According to experts, only 20 percent of the knowledge physicians use to make diagnosis and treatment decisions today is evidence based on thoughts shared by Manzoor Farid (2014). It is also observed that one in five diagnoses are incorrect or incomplete and nearly 1.5 million medication errors are made in the United States every year. The complexity of medical decision making is growing day by day and addressing this challenge is an important issue in front of health care systems. Smart health care systems designed using technological advances can handle this issue intelligently. Krawiec et al. (2015) discussed a system prototype with focus on major three components for improving health outcomes, reducing cost and expanding access to care: short-term care planning, chronic disease management and home care, and population-based evidence creation. Bui, Nicola, and Michele Zorzi (2011) have proposed the Internet of Things communication framework for health care applications. In this paper, they describe how the Internet of Things can be the main enabler for distributed health care applications and, conversely, how health care is one of the most promising killer applications for the IoT. Catarinucci et al. (2015) have proposed IoT-aware architecture for smart health care systems. Lee and Byung Mun (2014) have discussed design requirements for IoT health care models using an open IoT platform. Amendola et al. (2014) have discussed RFID technology for IoT-based personal health care in smart spaces. But these kinds of systems are working as individual components and integrated solutions with personalisation are missing.

Kanitkar et al. (2019a, b, c) designed an automated system for the diagnosis of various cervical abnormalities like cervical cancer, ascus, metaplasia, erosion, inflammation, candidiasis and nabothian cysts.

Figure 4.5 shows the components of a typical smart e-health care system. A large amount of heterogeneous time-series data is generated through various

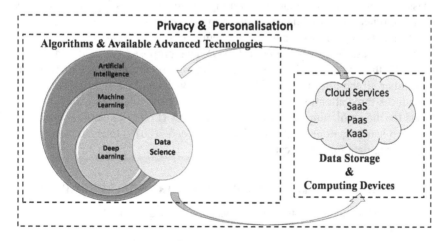

Figure 4.5 Components of smart e-health care system.

input sensors and interfaces. This data is stored on cloud systems and Hadoop clusters. and further transmitted for analysis and pattern extractions. Applications like glycaemia detection, cancer detection and skin disease detection requires in-depth analysis of images/video data. Advances in cloud services such as infrastructure as a service (IAAS) for delivering computing infrastructure as on-demand services, platform as a service (PAAS), which is a cloud delivery model for applications composed of services managed by the third party, and software as a service (SAAS) allowing the users to run existing online application ensures on-demand resource availability and helps designing the smart e-health care system at affordable prices. Design of smart systems involve artificial intelligence algorithms involving state-of-the-art data analytics and machine learning. Various variants of supervised, unsupervised and reinforcement learning allow the health care systems to identify the diseases and predict the future possibilities of the diseases. The large amounts of generated time-series data and available computational power allows the use of deep learning algorithms. In recent days, deep learning algorithms have shown great success in video, image and text data analysis with precision and speed, which otherwise was very challenging.

There are many different variations of the neural networks algorithms that use various substrategies within the domain.

Convolutional neural networks (CNN) have provided the best results in image-related applications that require the segmentation, classification and predictions of the possible anomalies if present in images. In the medical domain, the image data is generated through many sources like MRI, X-rays, CT scan, ultrasound, functional MRI, mammography or sonography. Kayalibay et al. (2017) have applied CNN and Kim (2015) deep CNN for the tasks such as image superresolution and imagenet classification (2012). Biomedical segmentation by O. Ronneberger (2015) and Szegedy (2015) and semantic segmentation by Long (2015) were also reported. Recurrent neural networks (RNNs) are a class of neural networks that are naturally suited for processing the time-series data and other sequential data as video clips of medical inspections. RNN is also used successfully for text analysis, but sometimes its performance degrades for long sentences. Fenglong Ma et al. (2017) have proposed the modified version of RNN, that is, dipole. It employs bidirectional recurrent neural networks to remember all the information of both the past visits and the future visits, and it introduces three attention mechanisms to measure the relationships of different visits for the prediction. Long short-term memory (LSTM) is another variant of the deep learning model that is used by Verma et al. (2019) to accurately predict the missing data for health care.

4.6 CHALLENGES AND SCOPE FOR FURTHER RESEARCH

The major challenges in e-health care system are discussed in this section.

- Security and privacy of large amount of personalised data

Data generated through various devices, such as Web and Android devices, that collect body parameters of the person are shared over the globe within seconds. Here, authorised and personalised access of data is the major challenge. Availability of recent advancement in blockchain technology is one of the solutions to address this problem.

- Intelligent computer vision and data analytics algorithms for automatic analysis of diseases from given data

This is especially helpful in remote places where very specialised medical services are unavailable.

- Availability of energy efficient sensors at affordable price, accurate with easy to wear

In order to make these smart e-health care systems more popular and to bring them in routine, we will have to ensure the easy and affordable availability of the related wearable devices / sensors. Affordability of the sensors and its accuracy have always been a challenge. This is one of the areas that needs the attention of researchers. Progress and achievements in sensors and wearable devices will ensure the growth of health care.

- Development of tools for accommodating rapid changes in e-health care system

Drastic changes in lifestyle and environmental changes demand robust and effective health care systems. During the COVID-19 period, we experienced this need. The countries which adopted changes rapidly and developed the solutions to tackle these problems were less affected by it or could successfully survive through it. Right things are adopted at the right time. The researchers made efforts for making products like low-cost, low-power-consuming ventilators, automatic mask detector systems, robots for taking care of patients, sanitisation kit and many more. Design engineers and researchers have a wide scope to make interface between human and technology. Using human–computer interaction (HCI), researchers have great goals and challenges to design user-centric interface for health care systems.

- Consistency and completeness in data

The effective use of electronic gadgets and Web applications is providing a huge amount of data for health care systems. The completeness and authenticity of data is a very important and challenging task for e-health care systems. The use of blockchain can be made for checking authenticity of data. There is a scope to design, implement and validate smart contracts for health care systems. These medical history and individual health status can be associated with a unique identification number of a person.

4.7 DISCUSSION AND CONCLUSION

Managing large amounts of heterogeneous data in the health care domain is very challenging. However, it provides the opportunity of in-depth analysis of diseases from patient history collected through various sensors. Again this collected data is not integrated and consistent for understanding the patient's history, so maintaining the consistent data is another challenge in the domain. Many times, the algorithms that have shown success on research data fail on real-time data. Worldwide many researchers are working on various problems independently. There is an intense need to bring the progress in this domain on the same platform to get unique and robust solutions. Here, the intention of providing robust solution is not to replace the medical experts or the doctors but intention to provide the sophisticated medical services at every corner of the world including developing and underdeveloped countries at affordable prices.

Even though various commercial and research organisations are working on designing health care systems, the domain is still away from the off-the-shelf technology and there are lots of research opportunities in the domain.

Bibliography

Acosta-Mesa, H. G., Rechy-Ramírez, F., Mezura-Montes, E., Cruz-Ramírez, N., and Jiménez, R. H. (2014). Application of time series discretisation using evolutionary programming for classification of precancerous cervical lesions. *Journal of Biomedical Informatics*, 49, 73–83.

Amendola, S., Lodato, R., Manzari, S., Occhiuzzi, C., and Marrocco, G. (2014). RFID technology for IoT-based personal healthcare in smart spaces. *IEEE Internet of Things Journal*, 1(2), 144–152.

Ayilara, O. F., Zhang, L., Sajobi, T. T., Sawatzky, R., Bohm, E., and Lix, L. M. (2019). Impact of missing data on bias and precision when estimating change in patient-reported outcomes from a clinical registry. *Health and Quality of Life Outcomes*, 17(1), 106.

Bui, N., and Zorzi, M. (2011). Health care applications: a solution based on the internet of things. In Proceedings of the 4th International Symposium on Applied Sciences in Biomedical and Communication Technologies, pp. 1–5.

Catarinucci, L., De Donno, D., Mainetti, L., Palano, L., Patrono, L., Stefanizzi, M. L., and Tarricone, L. (2015). An IoT-aware architecture for smart healthcare systems. *IEEE Internet of Things Journal*, 2(6), 515–526.

Duggal, R., Shukla, S., Chandra, S., Shukla, B., and Khatri, S. K. (2016). Impact of selected pre-processing techniques on prediction of risk of early readmission for diabetic patients in India. *International Journal of Diabetes in Developing Countries*, 36(4), 469–476.

Ebenuwa, S. H., Sharif, M. S., Alazab, M., and Al-Nemrat, A. (2019). Variance ranking attributes selection techniques for binary classification problems in imbalance data. *IEEE Access*, 7, 24649–24666.

Edemekong, A. P. F. and Haydel, M. J. (2018). Health Insurance Portability and Accountability Act (HIPAA). in StatPearls [Internet], StatPearls Publishing.

Islam, M.A., Talukder, M.H., and Hasan, M.M. (2013). Speckle noise reduction from ultrasound image using modified binning method and fuzzy inference system. In: 2013 2nd International Conference on Advances in Electrical Engineering (ICAEE), pp. 359–362.

Jain, D., and Singh, V. (2018). Feature selection and classification systems for chronic disease prediction: A review. *Egyptian Informatics Journal*, 19(3), 179–189.

Kanitkar, A., Joshi, V., Karwa, Y., Gindi, S., and Kale, G. V. (2019a, August). Comparison of machine Learning Algorithms for Cervical Abnormality Detection. In 2019 Twelfth International Conference on Contemporary Computing (IC3) (pp. 1–6). IEEE.

Kanitkar, A., Kulkarni, R., Joshi, V., Karwa, Y., Gindi, S., and Kale, G. (2019b, August). Automatic detection of cervical region from VIA and VILI images using machine learning. In 2019 IEEE International Conference on Computational Science and Engineering (CSE) and IEEE International Conference on Embedded and Ubiquitous Computing (EUC) (pp. 1–6). IEEE.

Kanitkar, A., Kulkarni, R., Joshi, V., Karwa, Y., Gindi, S., and Kale, G. V. (2019c, October). Automated system for cervical abnormality detection using machine learning. In 2019 IEEE 10th Annual Ubiquitous Computing, Electronics and Mobile Communication Conference (UEMCON) (pp. 0785–0790). IEEE.

Kayalibay, B., Jensen, G., and van der Smagt, P. (2017). CNN-based segmentation of medical imaging data. arXiv preprint arXiv:1701.03056.

Kim, J., Lee, J. K., and Lee, K. M. (2015). Accurate image super-resolution using very deep convolutional networks. arXiv:1511.04587v1 [cs.CV]

Krawiec, R. J., et al. (2015) How the IoT and patient-generated data can unlock health care value, Life Sciences and Health Care reports (deloitte.com)

Krizhevsky, A., Sutskever, I., and Hinton, G. E. (2012) Imagenet classification with deep convolutional neural networks. *Advances in Neural Information Processing Systems* 25(2), doi:10.1145/3065386

Lee, B. M. (2014). Design requirements for IoT healthcare model using an open IoT platform. computer, 4(5).

Long, J., Shelhammer, E., and Darrell, T. (2015). Fully convolutional networks for semantic segmentation. arXiv:1411.4038v2 [cs.CV].

Ma, F., Chitta, R., Zhou, J., You, Q., Sun, T., and Gao, J. (2017, August). Dipole: Diagnosis prediction in healthcare via attention-based bidirectional recurrent neural networks. In Proceedings of the 23rd ACM SIGKDD international conference on knowledge discovery and data mining (pp. 1903–1911).

Mezghani, E., Exposito, E., Drira, K., Da Silveira, M., and Pruski, C. (2015). A semantic big data platform for integrating heterogeneous wearable data in healthcare. *Journal of Medical Systems*, 39(12), 1–8.

Monrat, A.A., Schelén, O., and Andersson. K. (2019). A survey of blockchain from the perspectives of applications, challenges, and opportunities, *IEEE Access*, 7, 117134–117151.

Nosowsky, R., and Giordano, T. J. (2006). The Health Insurance Portability and Accountability Act of 1996 (HIPAA) privacy rule: Implications for clinical research. *Annual Review of Medicine*, 57, 575–590.

Panay, B., Baloian, N., Pino, J. A., Peñafiel, S., Sanson, H., and Bersano, N. (2019). Predicting health care costs using evidence regression. In Multidisciplinary Digital Publishing Institute Proceedings, Vol. 31, No. 1, p. 74.

Pathak, J., Bailey, K. R., Beebe, C. E., Bethard, S., Carrell, D. S., Chen, P. J., ... and Chute, C. G. (2013). Normalisation and standardisation of electronic health records for high-throughput phenotyping: the SHARPn consortium. *Journal of the American Medical Informatics Association*, 20(e2), e341–e348.

Peng, C., Goswami, P., and Bai, G. (2020). A literature review of current technologies on health data integration for patient-centered health management. *Health Informatics Journal*, 26(3), 1926-1951.

Ronneberger, O., Fischer, P., and Brox, T. (2015). U-net: Convolutional networks for bio-medical image segmentation. arXiv:1505.04597v1 [cs.CV].

Salem, N., Malik, H., and Shams, A. (2019). Medical image enhancement based on histo-gram algorithms. *Procedia Computer Science*, 163, 300–311.

Spencer, R., Thabtah, F., Abdelhamid, N., and Thompson, M. (2020). Exploring feature selection and classification methods for predicting heart disease. *Digital Health*, 6, https://doi.org/10.1177%2F2055207620914777

Szegedy, C., Liu, W., Jia, Y. et al. (2015). Going deeper with convolutions. In IEEE Conference on Computer Vision and Pattern Recognition, pp. 1–9.

Verma, H., and Kumar, S. (2019, January). An accurate missing data prediction method using LSTM based deep learning for health care. In Proceedings of the 20th International Conference on Distributed Computing and Networking (pp. 371–376).

Wanzala, M. N., Oloo, J. A., Nguka, G., and Were, V. (2019). Application of principal com-ponent analysis to assess health systems capacity using cross sectional data in rural western Kenya. *American Journal of Public Health*, 7(1), 27–32.

Wang, W., Hoang, D. T., Hu, P., Xiong, Z., Niyato, D., Wang, P., Wen, Y. and Kim, D. I. (2019).A survey on consensus mechanisms and mining strategy management in blockchain networks. *IEEE Access*, v7, 22328–22370, 2019.

Chapter 5

Comparative study on SMS spam message detection with different machine learning methods for safety communication

N. Krishnamoorthy, Suresh Muthusamy,
Seyedali Mirjalili, M. Vignesh, R. Vishnuhari, and
S.K. Sanjeev Raja

CONTENTS

5.1 INTRODUCTION

The steady rise in the use of mobile phones has led to an expansion of the telecommunication networks. One of the main factors driving the use of mobile phones has been the short message service (SMS). SMS is the distribution of small text messages from one device to other devices. Subscribers use SMS for personal and business purposes. They also receive spam messages frequently without knowing origin of the senders. This annoys the users or clients to the core. SMS spam has become a thorny issue for subscribers. Spammers have exploited the SMS and have been spamming the subscribers for promoting their business, products or services. Email spam filtering is easier than SMS spam filtering because email majorly contains formal words. Spam is an unsolicited message sent for the purpose of commercial advertising. These spam messages are usually sent by business

DOI: 10.1201/9781003335801-5

organisations in order to promote their products or services. The spam messages are sent to a large number of subscribers at the same time and to different places. Hackers are finding multiple ways to steal the subscriber data in recent times. SMS offers an uncomplicated way of hacking data by providing unsolicited information. This shows that spam threat is not geographically limited as it can go beyond the international borders. The links that are attached in the spam messages may lead into a non-secured connection that, in turn, may leak the subscriber's data. This poses a threat to user's privacy. The dataset related to SMS are rare in public websites as SMS services are run by private companies and they do not disclose customer data for research studies. We detect SMS spam messages using some existing machine learning algorithms and also compare which algorithm suits the best by considering accuracy as the main factor. This paper is organised into the following sections: literature review, system, outcomes and deliberations followed by conclusions.

5.2 LITERATURE SURVEY

Paras Sethi, Vaibhav Bhandari, and Bhavna Kohli (2017) conducted research on SMS spam recognition and evaluation of different machine learning methods. Stop words were removed from the data and most occurring tokens were displayed. Two important features, message length and count vector, were fed into the feature extraction. Information gain (IG) is used as attribute quality metric. Based on the experiments, naïve Bayes Classifier outperformed the remaining algorithms with an accuracy of 98.4 percent.

Amani Alzahrani and Danda B. Rawat (2019) performed a comparative study of existing machine learning algorithms for detecting the SMS spam messages. They used (1) naive Bayes classifier; (2) logistic regression; (3) support vector machines; and (4) neural networks. The SMS spam message dataset provided by UCI machine learning repository has been employed here. They considered accuracy as the primary evaluation metric and trained these existing models to find which algorithm is suitable to detect SMS spam messages. They concluded that neural network is the most effective model to detect SMS spam messages among the trained models.

Sahar Bo-saeed, Iyad Kateeb, and Rashid Mehmood (2020) performed a study titled 'A Fog Augmented Machine Learning based SMS Spam Detection and Classification System'. Five filters were combined and deployed to preprocess the data and extract the features. Lovins stemmer is used to remove the longest suffix from the words and for converting it into valid words. Further, a mobile application was built with a machine learning algorithm to detect the spam messages. Their evaluation concluded that naive Bayes algorithm provides the best method to find SMS spam messages as it outperforms the remaining algorithms.

Pavas Navaney, Gaurav Dubey, and Ajay Rana (2018) used supervised machine learning algorithms for SMS spam filtering. Initially the cleaning of the text messages in the dataset was done. The tokens were then represented through sparse matrix, which displays the dataset in the matrix format called document term matrix. Word Cloud was used to display the most frequent words. Based on the experiment, support vector machine classifier outperforms the remaining classifiers in terms of accuracy.

Krishnamoorthy, Kalaimagal, Shankar and Asif (2018) proposed a new IoT technique with the mobile communication to authenticate the status of the conventional device for security and safety issues.

Rohit Kumar Kaliyar, Pratik Narang and Anurag Goswami (2018) performed a study called 'SMS Spam Filtering on Multiple Background Datasets using Machine Learning Techniques'. The data was preprocessed by removing the stop words. They used Bag of Words for extracting the features and converting the words into a vector matrix. They used a mix instance-based dataset for training and testing the model. For the mix instance dataset, they obtained a good accuracy of around 95 percent.

Nurulhada Firdaus Mohd Azmi, Suriyati Chaprat, Haslina Md Sarkan, Yazriwati Yahya and Suriani Mohd Sam (2019) detected SMS spam messages using term frequency, inverse document frequency and random forest algorithm. In the preprocessing, they removed stop words and obtained a good number of recurrent words in both spam and ham messages. They then employed TF-IDF (term's frequency (TF) and its inverse document frequency (IDF)) for feature extraction. The extracted features were used for training and testing the model and compared with the performance of random forest classifier. Based on the tests, random forest algorithm gives more accuracy than other methods at 97.50 percent.

Saied Sheiki and Mohammed Taghi Kheirabadi (2020) sought to find an efficient form for SMS spam discovery using content-based attributes and average neural network. C# net framework was used for preprocessing and feature extraction. The proposed model's performance was evaluated using standard evaluation metrics. When accuracy was treated as the primary metric, average neural network provided an accuracy of 98.8 percent where their model resulted in an accuracy of 97.2 percent.

Mehul Gupta, Aditya Bakliwal, Shubhangi Agarwal and Pulkit Mehndiratta (2018) used two datasets of SMS spam collection (UCI) and Indian dataset with Hindi messages. After completing the data preprocessing like removing the stop words, Keras tokeniser was used for the tokenisation of words. Count vector followed by TF-IDF vectoriser was used to extract the features by creating the bag of words. Convolutional neural network was found to have a higher accuracy of 99.10 percent for dataset 1 and 98.25 percent for dataset 2.

5.3 METHODOLOGY

5.3.1 Preprocessing

Data preprocessing is very important in order to train the machine learning model better. We used the field names from 'v1' and 'v2' to 'label' and 'text' for better understanding of the dataset. The data label will be converted into a binary as machine learning model will be trained in numbers or binary instead of raw text.

5.3.1.1 Data preparation

1. Loading and cleaning the data to remove the punctuation and numbers.
2. Converting the entire alphabets in the words to lower case.
3. Removing stop words that does not contribute anything to the vocabulary.

Then Word Cloud shows the most frequent words in the data. The words with smaller font represents the least frequent words and the words with a larger font represents the most frequent words.

5.3.2 Feature extraction

Lemmatisation is the process of grouping words that have similar meanings to one. Lemmatisation brings context to the words and performs a morphological analysis of the words. Wordnet is publicly available philological database of over 200 languages that provides linguistic relationships between words. It is one of the earliest and most common used lemmatiser technique.

Bag of Words (BoW) model is the most significant technique used in feature extraction of the text data. Count Vectorizer creates a matrix in which each unique word in the text represented by the field and each instance is represented as a row. The value in each cell represents the count of the words in that text sample.

Figure 5.1 Word Cloud of spam messages.

Figure 5.2 Word Cloud of ham messages.

5.3.3 Training and testing the model

After extracting the features in the dataset, we have to train the model. The dataset is divided into 75 percent (4,179) for training set and remaining 25 percent (1,393) for testing set. For training and testing the model, the values in the field 'Type' is taken as dependent variable X and the text messages in the field 'Text' is taken as independent variable Y. We train the dataset with the four existing machine learning models.

1. The naive Bayes classifier creates a probabilistic model for cataloguing of SMS messages. This classifier assumes that the occurrence of a specific feature in a class is not linked to the existence of any other feature.
2. Logistic regression is a supervised machine learning algorithm adopted to forecast the probability of the target variable. Here, we take type as the target variable and predicting the type of the text message between ham and spam.
3. Decision tree classifier is a predictive modelling tool that can be functional across many areas. Decision trees can be developed by an algorithmic loom that can divide the dataset in diverse ways based on variety of circumstances.
4. Random forest algorithm creates multiple decision trees on data examples and merges them to predict the outcome and finally chooses the best result by the way of voting. The construction of trees is based on the random selection of the features.

5.3.4 Comparison and analysis of algorithms

After training the dataset with the existing models, we applied evaluation metrics for the models. The metrics used to find the suitable algorithm as follows: precision score, accuracy score and recall score. Accuracy will the primary factor for choosing the best suitable algorithm among these trained models for detecting the SMS spam messages.

5.3.5 Flow diagram

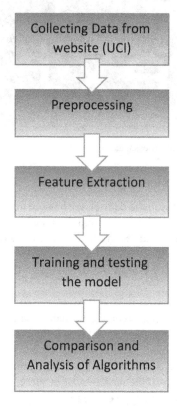

Figure 5.3 Representation of the proposed work.

5.4 RESULTS AND DISCUSSION

5.4.1 Dataset

The SMS spam collection is an open set of SMS labelled messages that have been composed for mobile phone spam investigation available in the University of California Irvine (UCI) Machine Learning Repository. The dataset contains 5,572 instances consisting of both ham and spam messages. Among the 5,572 instances, we are splitting the dataset into training and testing sets. We have taken 4,179 instances for the training set and 1,393 instances for the testing set.

Table 5.1 Statistics of dataset

Context	Number of messages
Training dataset	4179
Testing dataset	1393
Total	5572

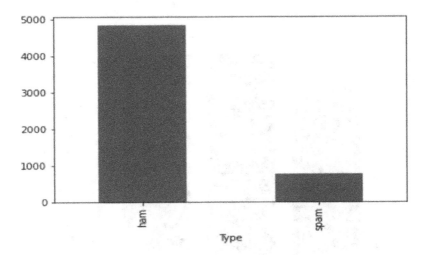

Figure 5.4 Statistics in the dataset.

5.4.2 Performance measure

The parameter used for evaluating the dataset is accuracy. Model accuracy is defined by the ability to predict both the positives and negatives out of all the predictions.

$$\text{Accuracy} = \frac{\text{Number of Correct Predictions}}{\text{Total Number of Predictions}} \tag{5.1}$$

For the binary classification within the matter of positives and negatives, accuracy are often computed as follows:

$$\text{Precision} = TP / (FP + TP) \tag{5.2}$$

$$\text{Recall} = TP / (FN + TP) \tag{5.3}$$

$$\text{Accuracy} = TP + TN / (TP + TN + FP + FN) \tag{5.4}$$

Table 5.2 Performance measures

Model	Precision	Recall	Accuracy
Naive Bayes	46.19	95.43	83.63
Random forest	100.00	87.31	98.21
Logistic regression	98.85	87.31	98.06
Decision tree	92.31	85.21	96.91

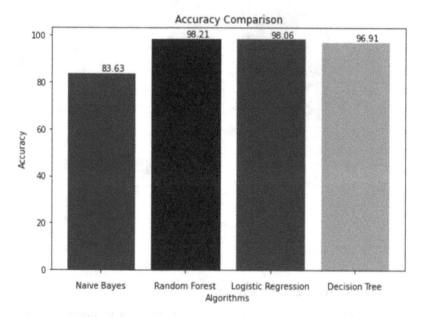

Figure 5.5 Comparison of algorithms in terms of accuracy.

5.5 CONCLUSION

Machine learning is the most popular technique used for the classification of SMS messages into spam and ham. It is one of the most successful techniques in filtering the email spams and it makes a preferable option for detecting SMS spam messages. The comparisons among four different algorithms were done. Accuracy was treated as the major factor or primary metric for selecting best algorithm for detecting SMS spam messages. The outcome obtained from our assessment of the algorithms demonstrate that random forest classifier outperforms the remaining three algorithms with accuracy score of 98.21 percent and logistic comes in terms of accuracy. Considering all the evaluation metrics that have been employed, we can easily conclude that random forest classifier is the most suitable machine learning algorithm to classify the SMS spam messages into spam or ham.

Bibliography

Alzahrani, A. and Rawat, D.B. (2019). Comparative Study of Machine Learning Algorithms for SMS Spam Detection. 2019 SoutheastCon, 1–6.

Bo-saeed, S., Katib, I.A. and Mehmood, R. (2020). A Fog-Augmented Machine Learning based SMS Spam Detection and Classification System. 2020 Fifth International Conference on Fog and Mobile Edge Computing (FMEC), pp. 325–330.

Gupta, M., Bakliwal, A., Agarwal, S. and Mehndiratta, P. (2018, August). A Comparative Study of Spam SMS Detection Using Machine Learning Classifiers. In 2018 Eleventh International Conference on Contemporary Computing (IC3) (pp. 1–7). IEEE.

Kaliyar, R. K., Narang, P. and Goswami, A. (2018, December). SMS Spam Filtering on Multiple Background Datasets Using Machine Learning Techniques: A Novel Approach. In 2018 IEEE 8th International Advance Computing Conference (IACC) (pp. 59–65). IEEE.

Krishnamoorthy, N., Kalaimagal, R., Shankar, S. G. and Asif, N. A. (2018). IoT-based Smart Door Locks. *International Journal on Future Revolution in Computer Science & Communication Engineering*, 4(3), 151–154.

Navaney, P., Dubey, G. and Rana, A. (2018, January). SMS Spam Filtering Using Supervised Machine Learning Algorithms. In 2018 8th International Conference on Cloud Computing, Data Science & Engineering (Confluence) (pp. 43–48). IEEE.

Sethi, P., Bhandari, V. and Kohli, B. (2017). SMS Spam Detection and Comparison of Various Machine Learning Algorithms. 2017 International Conference on Computing and Communication Technologies for Smart Nation (IC3TSN), pp. 28–31. doi: 10.1109/IC3TSN.2017.8284445.

Sheikhi, S., Kheirabadi, M. T. and Bazzazi, A. (2020). An Effective Model for SMS Spam Detection Using Content-based Features and Averaged Neural Network. *International Journal of Engineering*, 33(2), 221–228.

Sjarif, N. N. A., Azmi, N. F. M., Chuprat, S., Sarkan, H. M., Yahya, Y. and Sam, S. M. (2019). SMS Spam Message Detection Using Term Frequency-inverse Document Frequency and Random Forest Algorithm. *Procedia Computer Science*, 161, 509–515.

Chapter 6

An optimal more than one stage (MTOS) authentication model to ensure security in cloud computing

N. Krishnamoorthy, K. Nirmala Devi,
Suresh Muthusamy, Seyedali Mirjalili,
Essam H. Houssein, and Hitesh Panchal

CONTENTS

6.1 INTRODUCTION AND RELATED WORKS

Cloud computing enables the process of computation from simple laptops or desktops to across the globe through the internet. The various services such as infrastructure as one of the services, software as other service and platform as another service make the business very flexible for the real-world applications. The cloud computation process minimizes the cost and provides an optimal business data centre to the benefit of the business communities (Rittinghouse and Ransome, 2019). The infrastructure is provided based on demand by businesspeople and cost charged as per use. The cloud provides the service software to run the business as per the requirements mentioned by the user and provider, which provides services on a rental basis (Yousif et al., 2018). The platform is also provided as the service to deploy the needed software and to run the required application based on the demand as mentioned by the user. The hardware parts of the data centre will come under the service called infrastructure whereas the platform and software services will deal with the applications-deploying environment (Mell and Grance, 2011).

The major advantage of cloud computing is providing the clients with usage of the data or information and how the information is managed. Some of the major disadvantages as well as challenges in the cloud computing are data protection

DOI: 10.1201/9781003335801-6

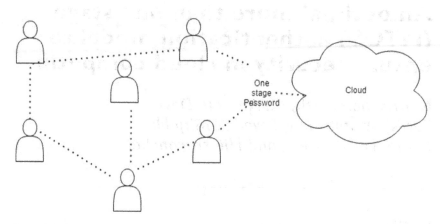

Figure 6.1 One stage authentication for cloud computing.

and information security (Xiao and Xiao, 2019; Kshetri, 2019; Jansen and Grance, 2020). The management of cloud authenticity among the cloud users plays a serious issue in the security of cloud computing. Various providers of the cloud utilize only one stage authentication approach as little as a word password for the clients to manage the services of the cloud as mentioned in Figure 6.1. Some of the cloud issuers use the fingerprint or eyeball scanning through a third-party authentication. But many restrictions arise for these approaches and one-stage authentication approach fails to perform.

Meanwhile, the major problem in cloud computing is how to enable an optimal methodology to perform authentication in order to provide data security and protect the secured information from illegal users. In this regard, the major objective of this proposed work is to provide an optimal model for protecting the information in the cloud, which expands the authentication system based on more than one-stage authentication. One-time password approach was introduced based on factors such as asymmetric scalar product preserving encryption and RSA to ensure authentication in cloud (Yassin, 2018). More than one stage authentication was implemented with some restrictions such as usage of same password at various levels, which utilize the services of the cloud (Dinesha and Agrawal, 2018). Another existing approach called tenant identification model was used to implement a secure cloud environment (Dinesha and Agrawal, 2018. An extended mutual authentication scheme was proposed to solve the issues in denial of service (DoS) attack and to protect the dynamic changing of password (Malar et al., 2020). The hill climbing heuristics are used for optimizing the scheduling. Simulations validate the proposed algorithm's performance, and results are evaluated by Makespan (Alsulaiman and El Saddik, 2019). Novel task assignment policies have been proposed by enhancing the hyper-heuristic approach for the type low task and high resource in the cloud environment (Krishnamoorthy and Asokan, 2014).

6.2 MORE THAN ONE STAGE SECURITY

More than one stage security is one of the branches of security containing various controls to provide security for the system. Information is protected from illegal users by utilizing the objects through a message authentication code (MAC) (Jaidhar, 2013). More than one stage security (MTOS) is used for major applications through a message authentication code, which concentrates only on highly secured information processed by the database, computer system and network (Faragallah et al., 2018). Each and every item present in more than one stage security is denoted as objects in which various levels of class of security are present. Each and every user is denoted as subject and it also contains various levels of class of security. In more than one stage security, an information plate is denoted as class l level of an object or subject Y and it is mentioned as IP(Y). More than one stage security contains various utilization controls in which Bell–LaPadulla plays a vital role and it has three rules. Rule one speaks about the utilization of read access of object O to the user Y when IP(Y) is greater than or equalized to IP(o). Rule two speaks about the utilization of write access of object O to the user Y when IP(Y) is smaller than or equalized to IP(o). Rule three speaks about the utilization of write access of object O to the user Y when IP(Y) is equalized to IP(o).

6.3 MORE THAN ONE AUTHENTICATION MODEL

The cloud data centre, which is located remotely, contains the databases of the client in which the security to those databases are provided by the cloud provider itself (Zhao et al., 2018). Normally the cloud providers make use of the more than one stage authentication model, which provides access to the data present in the cloud by the user in a secure way. Cloud computing uses graphical password, little word password, fingerprint authentication and third-party authentication. One-stage password for authentication will not provide you the proper security and there may be chances of attacks such as shoulder surfing attack, brute force attack and dictionary attack. If the malevolent users enter the network, all information regarding the genuine users may be accessed by the malevolent users (Sudha and Viswanatham, 2018). So it became a major threat or challenge for the cloud providers to protect the client's data from the malevolent users who enter to the network. So there is a need for an optimal secure cloud environment. In this regard, to solve the security issues in the cloud environment, the proposed system introduces an optimal more than one stage (MTOS) authentication model.

6.4 THE PROPOSED MORE THAN ONE STAGE (MTOS) AUTHENTICATION MODEL FOR CLOUD COMPUTING

The major objective of the proposed approach is to provide an optimal security for the information, infrastructure, platform and software for the cloud environment.

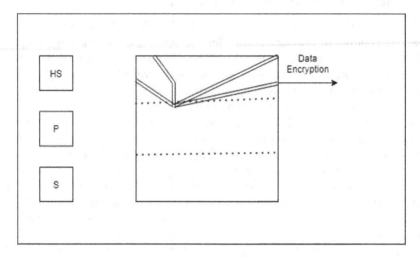

Figure 6.2 More than one stage (MTOS) authentication model.

The proposed approach is made up of a hybrid model, which includes the more than one-stage security and more than one-stage authentication. The proposed approach constitutes three stages of security and three stages of authentication from minimum to maximum. In this case, a single password, word password, is maintained in the minimum stage for the users, whereas in the second stage more than one password – word and fingerprint password – are used by the users. The secured data which is present in the third stage needs high level of protection, which is secured through three passwords for the users to enter and to download the information from the cloud environment. More than one stage (MTOS) model, which is the proposed approach, denotes the information location at various stages of security in which the information may occur in one location but may dynamically change its level of privilege and essentials. The hierarchy, which is denoted for the more than one stage model, contains security in three stages. The stages that are present are ranked from minimum to maximum as privilege (P), secret (S) and high secret (HS), which is shown in Figure 6.2. The client or a vendor who needs to utilize the information have to use the proper permission based on the characteristics of the stages.

The information is bisected into three stages based on information privilege at the P (Privilege) stage: The information available at the privilege level has minimum security in which even the genuine user uses only a single password to access the information present in the cloud environment. The client or vendor can enter the cloud to access the information by using the single password at this stage. The stage security has two verifications in which the user can only read the information present in stage C and the users in this stage are not permitted to edit in the minimum stage. In the High Secret stage, the information at the highest stage has the maximum level of security where the legal users access the information with three passwords. The information stored in the cloud will be encrypted before it can be used. The biometric password combined with RSA is used to extend the security of cloud privilege information.

The process of the proposed approach more than one stage (MTOS) model has been shown in Figure 6.3. The proposed model begins when a client enters

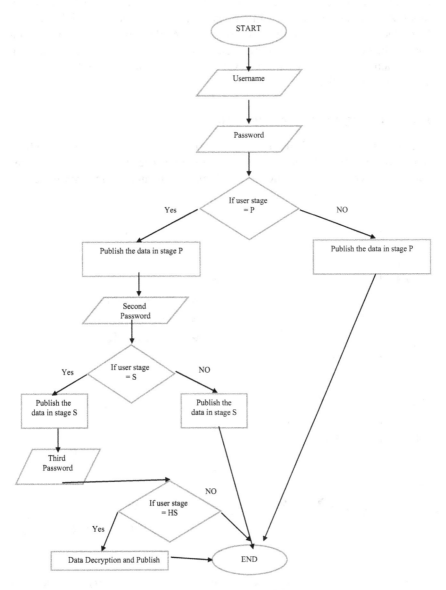

Figure 6.3 Flow diagram of the proposed approach MTOS.

his name and the primary secret key (literary secret phrase) as depicted in Figure 6.3. The framework confirms the client secret word, if the client level is equivalent to P, show the information at this level. In case the client's level is maximum than level P, at that point, the client needs to enter a biometric secret word. A similar cycle is accomplished for the client at level S by entering two passwords – printed and biometric. At that point, the plan confirms the client passwords, if the client level is equivalent to S, to show the information in this level and the lower level. In any case, if the client level is higher than S level, the client enters the third secret phrase. At that point, the proposed plot checks the client passwords, if the client level is equivalent to HS. Information is unscrambled and shown at this level and the lower levels of information are also accessible to the user.

6.5 RESULTS AND DISCUSSIONS

The proposed approach is framed according to the three stages of the authentication model, which depends on the user information. The proposed model utilizes various cases of passwords such as word, fingerprint and image password along

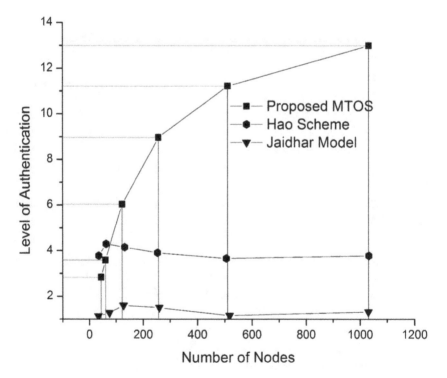

Figure 6.4 Comparison between existing and proposed system regarding level of authentication.

with cryptography technique. The proposed model, more than one stage authentication model has been compared with the some of the existing systems such as Hao Scheme and Jaidhar model. The experimental results are compared for the parameters such as percentage of security and level of authentication provided.

Figure 6.4 depicts the performance evaluation between the proposed MTOS model and the existing systems such as Hao scheme and the Jaidhar model. In the graph, the X-axis denotes the number of nodes and Y-axis denotes the level of authentication provided by the approaches. The graph clearly depicts the proposed system performs well and it provides the better authentication for maximum level of authentication when compared to the existing systems. After the fourth level, the proposed system performs remarkably well by providing maximum capacity of authentication in the cloud environment. Meanwhile, both the existing systems haven't performed well at the later maximum levels of authentication.

Figure 6.5 depicts the performance evaluation between the proposed MTOS model and the existing systems such as Hao scheme and the Jaidhar model. In the graph, the X-axis denotes the level of intruders and Y-axis denotes the percentage of security improved by the approaches. The proposed system MTOS improves

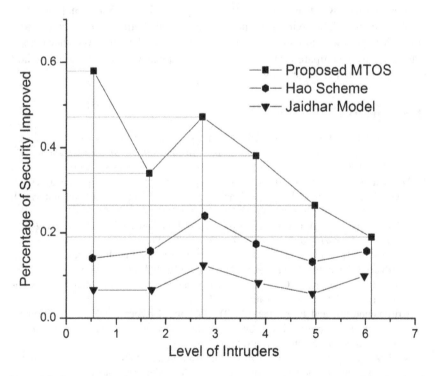

Figure 6.5 Comparison between existing and proposed system regarding level of confidentiality.

the security of the cloud environment between 0.2 and 0.6 percent whereas the existing systems improve the security of cloud environment below 0.2 percent. From the experimental results, it is clearly identified that the proposed system performs well compared to the existing system regarding the parameters, security and authentication.

6.6 CONCLUSION

Authentication presents significant difficulties for the security of distributed computing. Single-level validation has numerous issues for the most part with delicate information, as passwords are anything but difficult to break. The proposed method gave extra layer of security and provides to an answer for improving the verification framework dependent on staggered validation. The proposed plot comprises of three degrees of confirmation, and the information is divided into these levels depending upon affectability as Privilege (P), Secret (S) and High Secret (HS). Information at level P has the most minimal affectability. The client at this level has single printed secret key to get to this level information. The client at level (S) has two passwords – literary and biometrics secret key – to get to this level and the lower level. Client at level (HS) has three secret phrases – printed, biometrics secret word and picture sequencing secret key. The information at this level is the more delicate information so it is scrambled utilizing RSA calculation before it is put away in cloud data set. The proposed plan and two other best in class plans were compared. The underlying consequences of the proposed method were exceptionally encouraging.

Bibliography

F. A. Alsulaiman and A. El Saddik (2019). "Three-dimensional password for more secure authentication," *IEEE Transactions on Instrumentation and Measurement*, 57, 1929–1938.

H. Dinesha and V. Agrawal (2018). "Multi-level authentication technique for accessing cloud services," in 2012 International Conference on Computing, Communication and Applications, pp. 1–4.

O. S. Faragallah, E.-S. M. Ell, F. E. A. El-Samie, A. I. Sallam and H. S. El-Sayed (2018). *Multilevel Security for Relational Databases*. CRC Press.

C. Jaidhar (2013). "Enhanced mutual authentication scheme for cloud architecture," in Advance Computing Conference (IACC), 2013 IEEE 3rd International, pp. 70–75.

W. Jansen and T. Grance (2020). "Guidelines on security and privacy in public cloud computing," NIST Special Publication, vol. 800, pp. 10–11.

N. Krishnamoorthy, and R. Asokan (2014). "Optimized resource selection to promote grid scheduling using hill climbing algorithm." *International Journal of Computer Science and Telecommunications*, 5(2), 14–19.

N. Krishnamoorthy, K. Venkatachalam, R. Manikandan and P Prabu (in press). "A novel task assignment policies using enhanced hyper-heuristic approach in cloud." *International Journal of Cloud Computing*.

N. Kshetri (2019). "Privacy and security issues in cloud computing: The role of institutions and institutional evolution," *Telecommunications Policy*, 37, 372–386.

A. Malar, Christy Jeba, et al. (2020). "Multi constraints applied energy efficient routing technique based on ant colony optimization used for disaster resilient location detection in mobile ad-hoc network." *Journal of Ambient Intelligence and Humanized Computing*, 1–11.

P. Mell and T. Grance (2011). The NIST definition of cloud computing, https://nvlpubs. nist.gov/nistpubs/Legacy/SP/nistspecialpublication800-145.pdf

J. W. Rittinghouse and J. F. Ransome (2019). Cloud Computing: Implementation, Management, and Security. CRC Press.

S. Sudha and V. M. Viswanatham (2018). "Addressing security and privacy issues in cloud computing," *Journal of Theoretical and Applied Information Technology*, 48, 708–719.

Z. Xiao and Y. Xiao (2019). "Security and privacy in cloud computing," *Communications Surveys & Tutorials*, IEEE, 15, 843–859.

Yassin, A.A., Jin, H., Ibrahim, A., Qiang, W., Zou, D. (2013). Cloud Authentication Based on Anonymous One-Time Password. In: Han, YH., Park, DS., Jia, W., Yeo, SS. (eds) Ubiquitous Information Technologies and Applications. Lecture Notes in Electrical Engineering, vol 214. Springer, Dordrecht. https://doi.org/10.1007/978-94-007-5857-5_46

A. Yousif, M. Farouk and M. B. Bashir (2018), "A Cloud based framework for platform as a service," in Cloud Computing (ICCC), International Conference, 2018, 1–5. doi: 10.1109/CLOUDCOMP.2015.7149621.

H. Zhao, M. Xing, J. Zhao and H. Li (2018)., "Design and implementation of multi-level secure database management access control," *Journal of Applied Science and Engineering Innovation*, 2, 223–225.

Chapter 7

Robotic harvesters for strawberry and apple

Aman Mahore, Abhishek Patel, Mohit Kumar,
Rohit Nalawade, and Kanupriya Choudhary

CONTENTS

7.1 INTRODUCTION

India is the second largest producer of fruits (92 million tons in 2016–17) in the world followed by Brazil, the United States, Spain, Mexico, Italy and Indonesia (China is the first). Strawberry production in India was 0.5 million ton in the year 2015–16. Harvesting mature strawberries has been a difficult and time-consuming chore for a long time. Strawberry harvesting must be done by hand at a cool time of day to preserve the taste and look of the fruits (Qingchun et al., 2012). Strawberry production is heavily dependent on human labour, with working hours exceeding 20,000 working hours/ha, which is over 60 times higher than those for rice production in Japan (Shigehiko et al., 2010). From early in the morning, human harvesters must sit in a half-sitting position. Every year, harvesting work lasts around six months, and workers are overworked and exhausted due to poor working conditions. Working position not only affects the efficiency of harvesting

DOI: 10.1201/9781003335801-7

Table 7.1 Apple production in the different states of India (1,000 metric tons)

States/UTS	2012–13	2013–14	2014–15	2015–16	2016–17	2017–18
Jammu & Kashmir	1348.15	1647.69	1368.63	1672.72	1725.75	1808.33
Himachal Pradesh	412.40	738.72	625.20	777.13	468.13	446.57
Uttarakhand	123.23	77.45	106.10	61.94	62.06	58.66
Others	31.60	33.82	33.91	9.21	9.09	13.44
Total	1915.38	2497.68	2133.84	2521	2265	2327

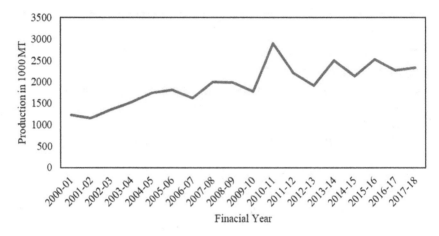

Figure 7.1 Trend of apple production in India (Anonymous, *Horticultural Statistics at a Glance*, 2018).

but also causes tiredness to the workers (Rajendra et al., 2009). Strawberry production in India was 0.5 million ton in the year 2015–16. Apple is one of the most widely cultivated tree fruits. India is the third largest producer of apple in the world having produced 2.9 million tons in the year 2015–16. Because of the taller fruit trees, harvesting must be done using ladders with steps, making manual harvesting unsafe and inefficient. Apple harvesting requires nearly 50–70 percent in terms of total hours of work. Still it is done manually in India.

7.2 PROBLEMS IN THE MANUAL HARVESTING OF APPLE AND STRAWBERRY

There are several problems associated with the manual harvesting of apple and strawberry. Some of these problems are as follows:

1. *Labour-intensive*: Every year, apple farmers must devote an average of eight hours per day to harvesting over the course of three months. Strawberry

production is heavily dependent on human labour, with working hours exceeding 20,000 working hours/ ha, which is over 60 times higher than those for rice production in Japan (Shigehiko et al., 2010).

2. *More skilled labourers are required:* Strawberry pericarp is so soft that the fruits must be harvested carefully by the workers to avoid damage. Workers must choose red and mature fruits among the many that have set. So, the proper skilled labours are required for harvesting of strawberries.

3. *Improper working posture*: Human workers must sit in a crouching or half-sitting position for the whole day to harvest strawberries. They have to bend to carry the fruits as well as to pick the fruits.

4. *Load and fatigue on workers*: Workers have to carry the harvested fruits to the end of the rows. Improper posture and load carried causes fatigue to the workers. The weight of the bucket for carrying apple may go over 25 kg, which they have to carry to the storage.

5. *Ladder required for tall trees*: Because apple trees are tall, harvesting requires the use of step ladders, making manual harvesting unsafe and inefficient.

6. *Lower efficiency and higher cost*: During manual harvesting of strawberries and apples, worker has to recognise the mature fruit, pick that mature fruit, keep that safely and carry that to the end of the row or carrier. Also, the improper working posture whether squatting, bending or standing on the ladder adds load and fatigue. Due to these reasons, the harvesting efficiency is much lower as well as higher harvesting cost is required in manual harvesting. Harvesting labour accounts for more than a quarter of the entire production cost.

As a result, there is a great demand to automate apple and strawberry harvesting. But there are certain challenges to overcome for the development of harvesting robots for strawberry and apple. These are as follows.

7.3 CHALLENGES FOR FRUIT HARVESTING ROBOT

- *Human-like perceptive capabilities*: It should have capability to differentiate between fruits and background. Also, it should be able to find the fruits behind the leaves and branches.
- *Avoiding collision*: It should not collide with branches and fruits to avoid the damage to the robot and the fruits.
- *High work efficiency and success rate*: Harvesting should be done in minimum time with highest accuracy. Also, the success rate of fruit recognition and picking should be maximum.
- *Variable lighting and environment conditions*: It should be capable of doing the job in a variable environment conditions such as varying levels of light and wind velocity.
- *Picking without damaging the fruits*: The skin of strawberry fruit is very soft and susceptible to damage. Apples are also susceptible to damage

and scratch due to mechanical harvesting. The taste, texture, colour and odour of the fruit should be retained after harvesting.

• *Assessment of maturity level*: Robots should be able to detect a high percentage of mature fruits. To achieve this goal, it is important to overcome basic existing difficulties such as occlusion and illumination variation. Robots must be able to differentiate between mature and immature fruits. Also, the detection of fruit cluster and background is required.

• *Calculation of fruit position*: Fruit position is the most important thing for harvesting. Position of the fruit in the space is required for the movement of the picking arm. It must be able to get the accurate 3D position of the fruit so the picking arm can hold the fruit in proper position to harvest.

7.4 STRAWBERRY HARVESTING ROBOTS

Strawberries are widely farmed all over the world due to their delicious taste and great production. In agricultural production, it is a highly successful and lucrative crop. Japan and China are the leading producers of strawberries. Due to labour insufficiencies and tedious work for human labour to harvest them spending hours in bending or squatting posture, some of the researchers developed robotic harvesters for strawberry.

Qingchun et al. constructed a harvesting robot to meet the requirements mentioned above. It mainly consists of six components.

7.4.1 Components of a strawberry harvester

1. *Four-wheel mobile platform*: The mobile platform is a four-wheel-drive vehicle, which makes the robot move with strong loading capacity and perfect mobility on the uneven terrain of the greenhouse at a speed of 0.3 m/s. Sonar sensors and a camera play the role of active perception device in the autonomous navigation. When the robot moves among the cultivation shelves, sonar sensors fixed on both sides of it will detect the distances between the vehicle and the iron-sheet on the shelves and help the control system to keep the vehicle moving forward in a straight line. The camera in the front of the mobile platform will capture the colour guideline on the ground and lead the robot to turn around when coming to the end of a crop line. The moving trajectory of the robot is shown in Figure 7.3. Due to the hybrid sensing strategy of sonar sensors and a video camera, the robot can move independently without other devices.

2. *6-DOF manipulator*: It is fixed on the vehicle 500 mm high from the ground and is used as an arm for positioning the end-effector during the harvesting operation. Its bottom joint can rotate in an angle range of 0–345° and the working space looks like a hemisphere with the diameter of 650 mm, which is feasible to handle the fruits on both sides of the robot. The manipulator has six degrees of freedom (6-DOF) and can carry 5 kg max-load (heavier than

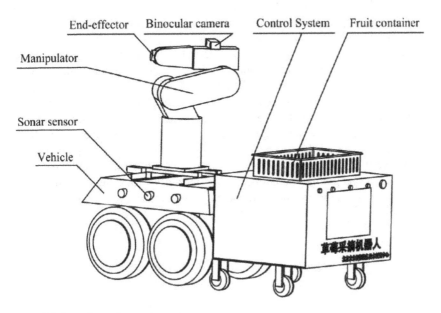

Figure 7.2 Strawberry harvesting robot.

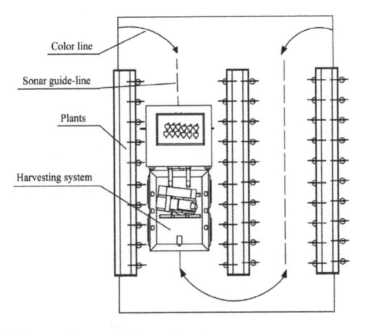

Figure 7.3 Moving trajectory of the harvesting robot.

our end-effector designed) with steady-state accuracy of ±0.02 mm. Besides the rotation-joint mode structure satisfies the requirement well for working in the limited space between crop and robot.

3. *Harvesting end-effector*: The end-effector has two fingers that create pneumatic gripper for grabbing peduncles, suction cup for holding fruit and electrical heating knife for peduncles cutting. As the end-effector nears the target fruit, the suction cup shaped as bellows is pushed out 100 mm to vacuum-hold the fruit, then pulled back to keep the peduncle between both fingers. The fingers are closed to hold the peduncle and the elastic layers of urethane are employed on the fingers to keep the peduncle safe from pinching off. Above the two fingers there are the electrical heating thread and the cutting board, respectively. The cord is heated to 200°C while the two fingers are closed to slice the peduncle. Heat-cutting destroys viruses on the cutting surface, preventing disease infection in plants.

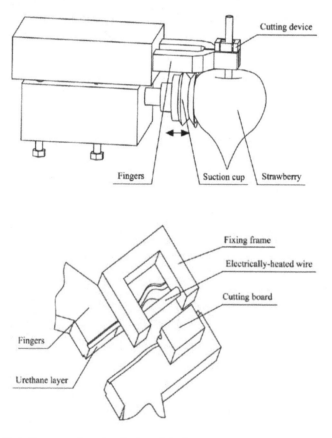

Figure 7.4 End effector and cutting device of the harvesting robot.

4. *Binocular vision unit*: The ripened strawberries are identified and located using the Point Grey Bumblebee2 binocular camera. The camera was installed on between the third and fourth joints of the manipulator, and acquired targets picture from each side at the right height, proving that the commercial vision unit is practical for decreasing the development cost while meeting the requirement of target detection. It has a focal length of 6 mm and a resolution of 1,024 (H) and 768 (V) pixels. The field view of the camera has 670 mm width and 500 mm height at 700 mm distance from camera to strawberries.

5. *Fruit container*: It is mounted over the robot to store the harvested strawberries in the container. When the harvesting tray is filled, it is directly returned to the stacker and next empty tray is produced. The tray is returned back to the stacker once the entire area has been harvested.

6. *Controller*: The target recognition, navigation, end-effector control processes and manipulator motion that make up the robotic harvesting task schedule are all handled by the system controller. A sensible schedule is essential for high performance and cooperative automatic operations.

7.4.2 Operational flow of harvesting

After startup and initialisation, the harvesting system will continue forward on its own due to telecontrol instructions transmitted by individuals. The navigation system will stop moving if the moving time reaches 2 s. The vision unit then starts identifying and locating mature strawberries in its view field, sending 3D location data of the accessible fruits to manipulator. With the operations of suck, grip and cut, the strawberry is separated from the plants by the end-effector, which is positioned to the target. Finally, the fruit will be carried and released over the container. A single picking cycle is now complete. After all of the ripe fruits on the left side have been picked, the manipulator's bottom joint will rotate 180 degrees to the right to pick fruits.

When both sides of the harvesting activity are completed, the robot will go to the front to collect other fruits until it receives telecontrol instructions to halt. When the manipulator returns to the initial gesture after complete harvesting cycle, if it is unsuccessful to separate the fruit from the plant due to position error or collision, it will still exist in same area of the field view. The robot will attempt to harvest another one five times before giving up.

7.4.3 Classification of strawberries

Rajendra et al. conducted a study for the development of algorithm for robot to harvest strawberries. They had classified strawberries in five different categories according to their surface visibility as follows.

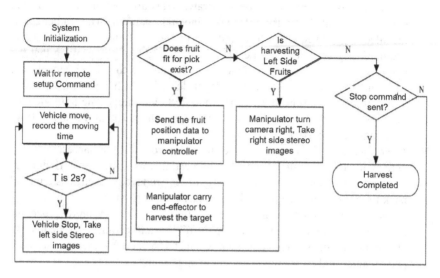

Figure 7.5 Task sequence of the harvesting operation.

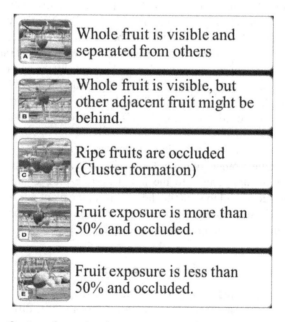

Figure 7.6 Classification of strawberries.

7.4.3.1 General flow of harvesting

(i) Robot looks for the red strawberry, which was converted using the HIS approach in this study.

(ii) Fruit's depth is calculated.

(iii) Fruit's maturity is determined, and if maturity>80 percent, the robot advances closer to the fruit.

(iv) Robot collects the image from the centre camera, correcting any differences between fruit and robot and computing the inclination angle of peduncle.

(v) Fruit is picked by rotating joint of wrist in the direction of peduncle. In this example, if there is no fruit in vacuum pad, robot will return to its starting position. If the fruit is successfully gathered, it is placed in the tray in a sequential order, and the robot returns to its starting position to harvest more strawberries.

1. *Colour conversion by the HSI method*: For colour conversion, a cylindrical model was employed. For the month of March, the following values were utilised to detect peduncles:

   $20 <$ Hue $< 45, 63 <$ Saturation $< 250, 80 <$ Intensity < 200

   The colour detection levels alter with the seasons. The hue of peduncles, for example, will be brownish green in May and June. To detect the colour of peduncles, the values must be changed in this scenario.

2. *Depth calculation*: The following equation is used for depth calculation:

$$f(d) = \frac{\mu \times F}{Disperity \times \varepsilon}$$

   Disparity and ε are inversely proportional to depth. To get the optimum depth, first compute the disparity, then multiply it by the optimised factor to get the corrected disparity. The golden section method was used to optimise the factor. We thought that the best value for factor ε is somewhere between 0.0 and 0.5.

3. *Maturity calculation*: First, 100 Type 'A' fruits of different maturity levels were selected. Second, the maturity levels of strawberries were decided by humans. Third, the same strawberry was kept in front of the machine to calculate the maturity. In this study, maturity of the fruit is calculated using the left camera. The equation used for maturity calculation is shown below:

where, $A = \sum_{i=1}^{i=p} R_i$ and $B = \sum_{i=1}^{i=p} R_i + \sum_{i=1}^{i=q} U_i$

R = Red Pixels, U = non-red pixels, p = total red pixels and q = total non-red pixels.

$$M = \frac{A}{B} \times 100$$

Here, M = maturity, percent, A = red pixels area in the fruit, B = total pixels of the fruit.

4. *Cluster detection*: Most of the strawberries during the experiment were found to be cluster type. Type C, D and E strawberries are known to the machine as single fruits. In this study, one method based on the gradient of the vectors was applied to detect the C type strawberries. For non-linear functions, the gradient vector at points (X_1, X_2) and $(\dot{X}_1, \dot{X}_2)$ were calculated as follows:

The convexity breakout of the external boundary at point (X_1, X_2) was checked using the following equation:

$$f\left(X_1, X_2\right) = \nabla f\left(\dot{X}_1, \dot{X}_2\right) \cdot \nabla f\left(\bar{X}_1, \bar{X}_2\right)$$

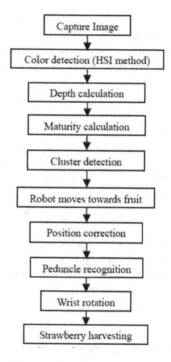

Figure 7.7 General flow of the harvesting process.

where, $\dot{X}_1 = X_1 - T$ and $\dot{X}_2 = X_2 - \varsigma$, T and ς are the arbitrary constants with small values.

If $(X_1, X_2) < 0$, The observed region of interest possesses convexity breakout.

5. *Position correction*: This was accomplished by taking the centre camera image at a distance of 100 mm from fruit target, which improved deviation. They analysed the image and found the strawberry that was closest to image's centre. Following the implementation of this strategy, we discovered that fruits could be harvested with a deviational error of ±5mm.

6. *Peduncle recognition*: The strawberry region is first recognised, and then a fret diameter restriction is applied on the item in the top portion of region. The peduncle is considered if the fret diameter is less than 0.6; otherwise, the existent object is leaf, immature fruits, strawberry stems or leaf stems.

   However, under some conditions, the proposed approach fails. The calyx is strongly curled if the two peduncles are joined, or if a little green strawberry is associated with the peduncle.

7.4.4 Maturity calculation

In Rajendra et al (2009)'s investigation, 100 strawberries were selected at random to better comprehend the machine's maturity detection. The strawberry was first examined by humans, and then the machine was presented the identical strawberry to judge its maturity. The link between maturity detected by humans and machine is depicted in Figure 7.8. It has been discovered that the estimate of maturity is influenced by two key aspects: (i) different surface visibility of the same fruit and (ii) distribution of the light source.

**Successful recognition
of peduncle**   **Peduncle is not visible**

Figure 7.8 Peduncle recognition.

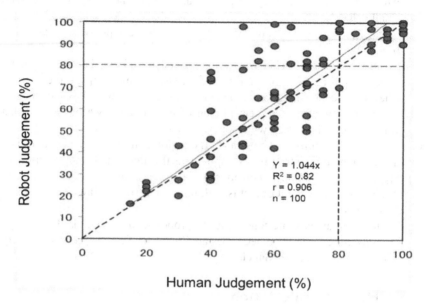

Figure 7.9 Comparison of maturity by humans and the machine.

The fruits above 80 percent maturity is considered as mature. Robot had recognised larger range of strawberries than human judgement. It is due to the narrow band of red colour, which causes the recognition of pinkish strawberries as mature ones. As a result, red colour band width should be enlarged to capture pinkish region of strawberry, while the green colour band should be used to capture the unripe area.

7.4.5 Success rate of peduncle recognition

As discussed earlier, the five different types of strawberries were classified. They had conducted a study for the recognition of peduncle in the three different months and the result is shown in Table 7.2.

The results clearly show that the recognition of single strawberry (A) was very high. But the recognition of strawberry having exposure below 50 percent (E) was found to be zero.

Shigehiko et al. (2010) had conducted study on harvesting of strawberry with suction picking and non-suction picking of strawberry. In the suction picking, vacuum pressure is applied to the target fruit and compressed air is blown to keep other adjoining fruits away. This way the error in position calculation and obstacles in between are avoided.

Table 7.2 Peduncle recognition success rate

Month	Type of fruits				
	A	B	C	D	E
March	91%	72%	57%	48%	0%
April	95%	77%	56%	69%	0%
June	71%	73%	60%	60%	0%

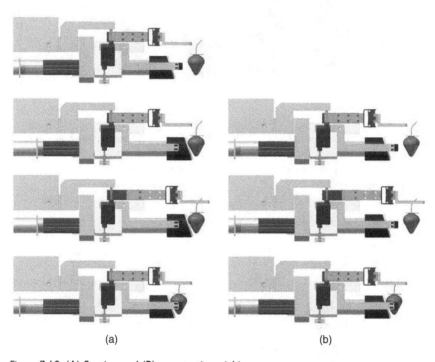

(a)   (b)

Figure 7.10 (A) Suction and (B) non-suction picking.

Figure 7.10 shows the comparison between suction and non-suction picking of strawberry. It clearly indicates that the successful harvest in tray is 72 percent in case of suction picking against 62 percent in non-suction picking.

During harvesting tests in the field, overall spontaneously hypertensives (SHRs) were 41.3 percent for suction picking and 34.9 percent for picking without suction, but they were still far from the numerical target of 60 percent. The execution time for harvesting fruit with its placement in tray, was 11.5 s, which is 2.5–3 times more than time required for manual harvesting (Shigehiko et al., 2010).

(A) Suction picking

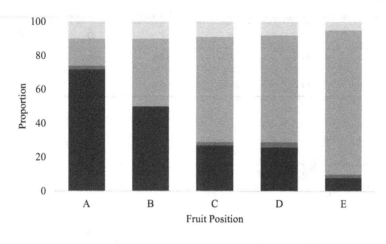

(B) Non- suction picking

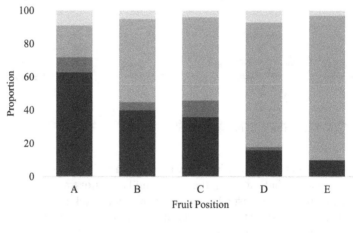

Figure 7.11 Comparison of the harvesting performance of suction picking and non-suction picking operations.

Table 7.3 Execution time of each motion

Operation	Avg execution time, s
Picking fruit	7.7
Placement into the way	3.8
Preparation of the tray	16.5
Changing the tray	15.0
Travelling for 200 mm	1.0

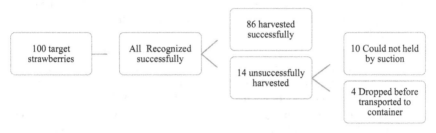

Figure 7.12 Results of strawberry harvesting with robot (Qingchun et al., 2012).

Qingchun et al. (2012) had built a robot based on the sonar-camera sensor, an autonomous navigation system and a six-degree-of-freedom manipulator. In the harvesting test, all 100 mature strawberry targets were automatically detected, according to the results. The success rate of harvesting was 86 percent, took 31.3 s on average, including single 10 s harvest operation. Fruit placement error was smaller than 4.6 mm on average.

7.5 APPLE HARVESTING ROBOT

The apple tree is a member of the rose family, and it is a delicious, pomaceous and most famous fruit. It is grown as a fruit tree all over the world. The apple tree is a deciduous tree that grows to a height of 1.8–4.6 m (6–15 feet) under cultivation and up to 12 m (39 feet) in the wild. Harvesting is one of the problems in the tall trees as it requires a larger number of labourers and ladders to reach there. So, the apple harvesting robot can be one of solution to the problems.

Zhao et al. (2016) created a robotic system with a manipulator, end-effector, and image-based vision servo control system for apple harvesting. The 5 DOF PRRRP structure of the manipulator was tuned geometrically to offer quasi-linear behaviour to simplify control technique. The spoon-shaped end-effector with pneumatically activated gripper was created to meet the needs of apple picking. Using a vision-based module, the harvesting robot did its work autonomously.

7.5.1 Major components of an apple harvesting robot

The apple harvesting robot prototype was created with efficiency along with cost effectiveness in mind. It mainly comprises of an autonomous vehicle, manipulator with five degrees of freedom (DOF), an end-effector, sensors, a vision system and a control system.

1. *The autonomous mobile vehicle*: Crawler-type mobile platform was selected as the mobile vehicle. It carried the power supplies, pneumatic pump, electronic hardware for data acquisition and control and the manipulator with the end-effector for cutting the fruit. The global positioning system (GPS) technology was used for autonomous navigation of the mobile vehicle, whose typical speed was 1.5 m/s.

2. *The manipulator*: Manipulator of harvesting robot with 5 DOF prismatic revolute-revolute-revolute-prismatic (PRRRP) structure to be mounted on autonomous mobile vehicle was designed. The first DOF was used for uplifting the whole manipulator. The middle three DOFs were for rotation, among which, the second driving arm was designed to rotate around the waist, and the third and fourth ones were rotation axes to move the terminal operator up and down. This DOF allowed the end-effector to move towards an arbitrary direction in the workspace. The fifth, and last, DOF was flexible and used for

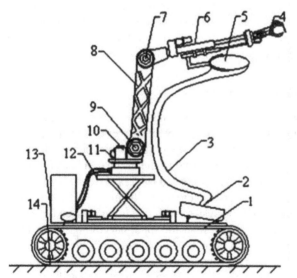

1. Mobile vehicle
2. Basket
3. Flexible band
4. End-effector
5. Gathering unit
6. Electric handspike
7. Minor arm motor
8. Major arm
9. Major arm motor
10. Waist motor
11. Waist
12. Lifting platform
13. Power and control equipment

Figure 7.13 Apple harvesting robots.

elongation, which made the end-effector reach the target location according to the robot control commands, thus achieving the harvesting of fruit.

3. *The end-effector*: A spoon-shaped end-effector (shown in Figure 7.3) is designed according to biological characteristics of spherical fruit, which are picked by means of cutting off the stalk. The end-effector contained the following parts: a gripper to grasp the fruit and an electric cutting device to separate the fruit from the stalk. The opening and closing of end-effector gripper was determined by some pneumatic devices, whose quick action, fast response characteristics were suitable for the switching control of the end-effector.

4. *The sensors on- end-effector*: The layout of sensors on end-effector, which includes a vision sensor, a position sensor, a collision sensor and a pressure sensor, is shown in Figure 7.4. The vision sensor, which uses high pixel colour charge coupled devices (CCD) video camera with universal serial bus (USB) interface and the video for windows (VFW) capture technology to form image acquisition system, plays a key role in completing image acquisition, fruit search and recognition. To obtain a wide visible field and not influenced by end-effector, the position of the vision sensor is in an eye-in-hand mode. The photoelectric sensors are used in pair. They were used to determine the fruit position with respect to the end-effector position. The collision sensor was used for obstacle avoidance during the process of harvesting. The function of pressure sensor is to determine the holding pressure of fruit. It helps in avoiding the damage to the fruit during harvesting.

5. *The control system*: At the centre of the control system was the host computer, which integrates the control interface and all of software modules to control the whole system. The sensor signal acquisition system and image acquisition system constituted the input section, which was used to collect external environment information for the harvesting robot. The output section included a servo driven motor, air pump and end-effector.

6. *The vision systems*: Recognition is the process of distinguishing the object of attention (the fruit) from the background (leaves, branches, sky and dirt). Segmentation is a type of image processing technique. When the colour of the fruit differs from its surroundings, a colour video camera is frequently used to identify it. Colour is a useful descriptor for enhancing an object in an image, making it easier to identify and extract objects from it. A colour charge-coupled device camera was used to capture apple photos, and a personal computer was used to analyse the images in order to recognise and locate the fruit. The red colour difference was used to improve the fruit image because fruit had largest red colour difference among objects in image (Bulanon et al. 2002).

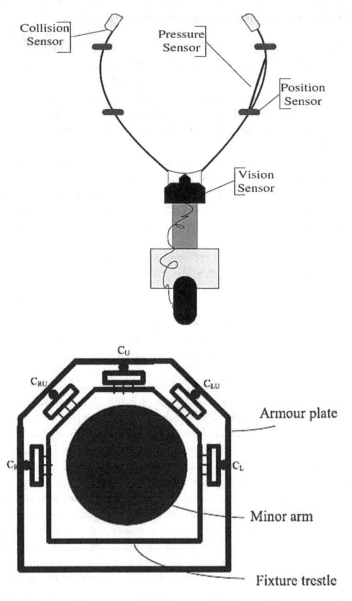

Figure 7.14 (A) Layout of sensors on-end-effector and (B) layout of sensors on-minor-arm.

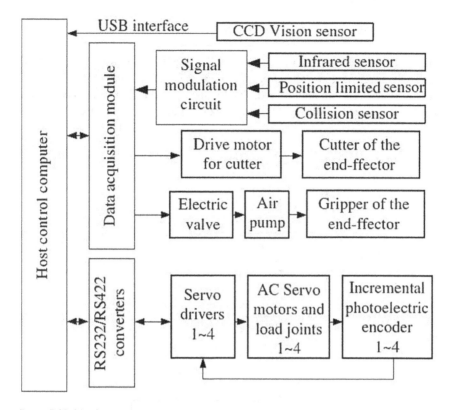

Figure 7.15 Hardware structure of apple harvesting robot control system (Zhao et al., 2016).

7.5.2 Apple recognition tests in different conditions

Bulanon et al. (2002) conducted apple recognition in four different natural lighting conditions:

a. Full sun, front-lighting
b. Full sun, back-lighting
c. Full sun, fruits in shade
d. Cloudy conditions

The colour properties used in this model are: luminance, Y; red colour, difference CR; green colour difference, CG; and blue colour difference, CB. The equations used to convert RGB values to luminance and colour difference signals are as follows:

$$Y = 0.3 R + 0.6 G + 0.1 B$$

$$C_R = R - Y, C_G = G - Y \ \& \ C_B = B - Y$$

where, R, G, and B are the red, green and blue colour intensity levels, which ranged between 0 and 255. Only brightness and red colour difference were employed because Fuji apple is red. The luminance and colour difference (LCD) space was created by plotting luminance against red colour difference.

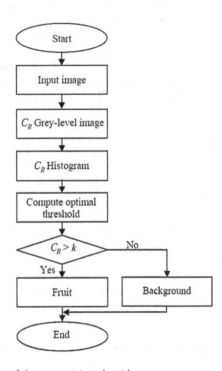

Figure 7.16 Flowchart of the recognition algorithm.

where C_R is red colour difference and k is optimal threshold. After the red colour different calculation, C_R, a histogram is made as shown in Figure 7.17. The lowest point in the graph shows the optimal threshold value.

The fruit has a bigger red colour difference than background in all the lighting circumstances, as seen by the spaces. As a result, red colour difference makes it easier to distinguish fruit from other two parts than luminance gap.

The thresholds utilised to segment photos under different illumination circumstances are varied, as shown in the Table 7.4. The images with the highest threshold values were those in front-lighting settings, followed by those in cloudy situations, fruits in shade and back-lighting conditions. This means that algorithm adapted to the type of condition on its own.

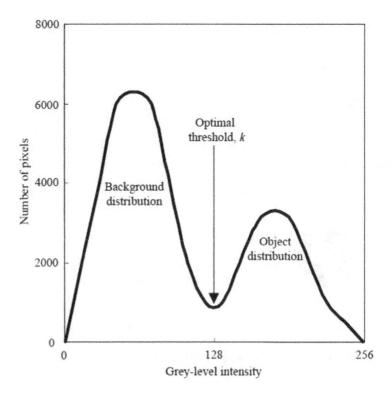

Figure 7.17 Example of bimodal histogram.

Table 7.4 Performance of the recognition algorithm (Bulanon et al., 2002)

Condition	Success, %	Error, %
Full sun, front lighting	92.2	1.5
Full sun, back lighting	89.1	18.2
Full sun, fruit in shade	88.6	1.3
Cloudy	95.4	0.9

Zhao et al. (2016) performed laboratory and field tests for the recognition and harvesting of the apples. The picking success rate was observed 77 percent and mean picking time was 15.4 s.

Original colour image of apple

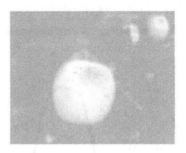

Enhanced image using red colour difference

Image after thresholding

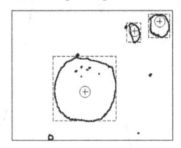

Image after edge and feature extraction

Figure 7.18 Apple recognition using optimal threshold.

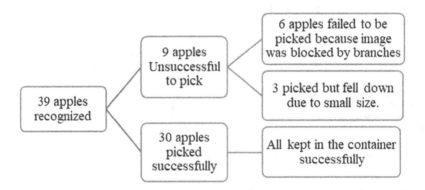

Figure 7.19 Results of apple harvesting robot (Zhao et al., 2016).

7.6 CONCLUSION

Harvesting of apple and strawberry using robots is challenge for the researchers. The most difficult part in this field is the recognition of the fruits in the tree and their maturity calculation. After that, the determination of the position of fruits in the tree from the manipulator and end-effector requires mathematical and imaging techniques, which are explained in the studies. Many of the works are successful in the initial phase of the recognition in the laboratory studies, which encourages the researchers to work for real-time challenges.

For the robot's practicability and commercialisation, future research should concentrate on the following three aspects:

(1) Reduced computation by optimising existing software programmes and techniques. At the same time, boosting the robot's practicality by improving the accuracy and speed of picking for obstructed or swinging fruits.

(2) Given the intricacy and uncertain nature of the working environment, further research into real-time obstacle avoidance and picking success should be conducted.

(3) Enhancing the configuration by upgrading the robot's mechanical structure to achieve all-purposes of robot, for example, other systems might be used to replace the manipulator and end-effector, or different freedom degrees could be used to pick fruit of various shapes and sizes. This would boost its adaptability, cut its overall cost, and help it become more commercial.

Bibliography

Anonymous. (2017). *AGROBOT Strawberry Harvester with Industrial Sensors*. Retrieved August 26, 2017, from Berry Picking at Its Best with Sensor Technology: www.pepp erl-fuchs.com/global/en/27566.htm

Anonymous. (2018). *Horticultural Statistics at a Glance 2018.* New Delhi: Horticulture Statistics Division.

Ashwini, K. (2016, January). Survey paper on fruit picking robots. *International Journal of Computer Science and Mobile Computing, 5*(1), 96–101.

Bachche, S. (2015). Deliberation on design strategies of automatic harvesting systems: A survey. *Robotics, 4*, 194–222.

Bakhtiari, A. A., & Hematian, A. (2013). Design, fabrication and evaluation of a picking mechanism for fruit harvesting. *Indian Journal of Agricultural Sciences, 83*(10), 27–30.

Bulanon, D., Kataoka, T., Ota, Y., & Hiroma, T. (2002). A segmentation algorithm for the automatic recognition of Fuji apples at harvest. *Biosystems Engineering, 83*(4), 405–412.

Bulanon, D., Kataoka, T., Ukamoto, H., & Hata, S. (2004). Development of a real-time machine vision system for the apple harvesting robot. *SICE Annual Conference.* Sapporo: Hokkaido Institute of Technology, Japan.

Cao, Q., Nagata, M., & Gejima, Y. (2000). Basic study on strawberry harvesting robot (part I) – Algorithm for locating and feature extracting of strawberry fruits. *Bio-Robotics, Information Technology and Intelligent.* Osaka, Japan. doi:10.1016/s1474-6670(17)36750-2

De-An, Z., Jidong, L., Wei, J., Ying, Z., & Yu, C. (2011). Design and control of an apple harvesting robot. *Biosystems Engineering, 110*, 112–122.

Earle-Richardson, G., Jenkins, P., Fulmer, S., Mason, C., Burdick, P., & May, J. (2005). An ergonomic intervention to reduce back strain among apple harvest workers in New York State. *Applied Ergonomics, 36*, 327–334.

Fernández, R., Salinas, C., Montes, H., & Sarria, J. (2014). Multisensory system for fruit harvesting robots. Experimental testing in natural scenarios and with different kinds of crops. *Sensors, 14*, 23886–23904.

Glavan, D., Babanatsas, T., & Merce, R. M. (2016). Study of harvesting methods and necessity of olive harvesting robot. *Annals of Faculty Engineering Hunedoara – International Journal of Engineering, 3*(August), 143–146.

Hayashi, S., Shigematsu, K., Yamamoto, S., Kobayashi, K., Kohno, Y., Kamata, J., & Kurita, M. (2010). Evaluation of a strawberry-harvesting robot in a field test. *Biosystems Engineering, 105*, 160–171.

Hayashi, S., Yamamoto, S., Tsubota, S., Y. O., K. K., Kamata, J., ... Peter, R. (2014). Automation technologies for strawberry harvesting and packing operations in Japan. *Journal of Berry Research, 4*, 19–27.

Ji, W., Zhao, D., Cheng, F., Xu, B., Zhang, Y., & Wang, J. (2012). Automatic recognition vision system guided for apple harvesting robot. *Computers and Electrical Engineering, 38*, 1186–1195.

Kushwaha, H. L., Sinha, J. P., Khura, T. K., Kushwaha, D. K., Ekka, U., Purushottam, M., & Singh, N. (2016). Status and scope of robotics in agriculture. *International Conference on Emerging Technologies in Agriculture and Food Engineering* (pp. 264–277). Kharagpur: Agricultural and Food Engineering Department, IIT Kharagpur.

Qingchun, F., Xiu, W., Wengang, Z., Quan, Q., & Kai, J. (2012, June). A new strawberry harvesting robot for elevated-trough culture. *International Journal of Agricultural and Biological Engineering, 5*(2), 1–8.

Rajendra, P., Kondo, N., Ninomiya, K., Kamata, J., Kurita, M., Shigi, T., ... Kohno, Y. (2009). Machine vision algorithm for robots to harvest strawberries in tabletop culture greenhouses. *Engineering in Agriculture, Environment and Food, 2*(1), 24–30.

Shigehiko et al., (2010). Sodium bicarbonate for prevention of contrast-induced acute kidney injury: a systematic review and meta-analysis. *Nephrology Dialysis Transplantation, 25*(3), 747–758. https://doi.org/10.1093/ndt/gfp389

Yamamoto, S., Hayashi, S., Yoshida, H., & Kobayashi, K. (2014). Development of a stationary robotic strawberry harvester with a picking mechanism that approaches the target fruit from below. *Japan Agricultural Research Quarterly, 48*(3), 261–269.

Yougman, A. (2017, February 7). Robot trials aim to cut strawberry harvesting time and control costs. Retrieved from www.producebusinessuk.com

Zhao, Y., Gong, L., Huang, Y., & Liu, C. (2016). A review of key techniques of vision-based control for harvesting robot. *Computers and Electronics in Agriculture, 127*, 311–323.

Chapter 8

Scalability and security requirements for the Internet of Things architecture

P. Deivendran, S. Ilaiyaraja, S. Selvakanmani, and K.S. Raghuram

CONTENTS

DOI: 10.1201/9781003335801-8

8.1 INTRODUCTION

The Internet of Things (IoT) has proven to be quite beneficial in terms of enhancing robotics, sufficiency, and customer comfort. The organisation that uses hardware, scheduling, sensors, and organisational networks to enable actual devices, vehicles, assemblies, and additives to collect and provide information is known as the Internet of Things (IoT). The data confirmed by IoT instruments is far from a range of sectors that have become vital during a range of applications such as a startup. The IoT environment is still in its infancy and can be stifled by programmers. Assuming we can deal with this knowledge and snoop gadgets using moderate device monitoring and firmware flaws.

Based on this understanding, numerous experts have proposed a blockchain arrangement to improve the degree of confirmation and discovery (Dalipi and Yayilgan, 2016), as well as the encryption and transfer of information (Dalipi and Yayilgan, 2016; Krylovskiy et al., 2015). These concepts, however, are dangerous in the face of digital attacks by programmers who use malware and device weaknesses to achieve their purpose. Furthermore, these linked devices can be employed and watched by programmers and cyber thieves to develop perplexing digital attacks, which can result in uncertain and lethal punishments. Malware recently utilised IoT devices and specified domain name service (DNS) agents to produce complex data. This distributed denial of service (DDoS) assault impacted the ISPs of many clients, including Netflix, GitHub, and Reddit, causing huge losses for these companies. However, these ideas are not assured in the case of digital attacks by programmers seeking to exploit malware and device flaws.

This research looks at the control of AI technologies as well as the viability of blockchain. The invention is defined by gradual isolation, data dispersion, and security of weak, hostile IoT equipment. For its IoT-strained design similarities, we combine using an approved blockchain for space and store IoT-gadget data where certain choices (Krylovskiy et al., 2015) and (Gao et al., 2018) in particular the linked devices are not their own. Limited calculation control contributes to the cycle and dynamic of evacuation.

8.1.1 Importance of scalability

An IoT framework links several sensors, powertrains and other online gadgets to allow data sharing and a larger variety of uses. This calls for strategic and framework development to address climate change and requirements of adaptable and flexible individuals. Variation refers to their capacity to respond and resolve when difficulties occur. The fundamental goal of making the device flexible is to meet changing demands, and it can never be static because people's interests and tastes change over time just as organically. It will increase weight, productivity and quality. It is necessary. Adaptability is important because it enables the frame to function smoothly or rapidly using accessible resources. When the memory demand for the framework grows, as information measurement in an adaptive framework rises, it does not develop to unacceptable levels at this stage (Ahmad et al., 2019). Either big or little, the gadget can function easily and swiftly. Therefore, it is important for a device to be flexible so that it is occasionally productive.

8.1.2 Vertical scalability

It is also known as the extension of existing machines or programmes, which can be enlarged by adding additional assets to them. To speed up operations, we strengthen the employee's handling skills. Moreover, by greatly increasing the preparation, primary memory, storage and organisational interface of this centre, we can scale the framework higher to satisfy further demands per frame. To facilitate administrative organisations, the number of processors increases. It relates, for example, to adding a CPU or RAM to a computer to one facility in a framework. This scalability of the present architecture allows them to make more cost-effective use of virtualisation innovation. Vertical adaptation is mainly used because it uses less power than supervising several employees and minimises managers' efforts, because one framework has only to be managed and monitored. Moreover, it is easier to execute, cost of development is lowered and it is a common programme. This kind of scalability has advantages and disadvantages, involving an increased chance of equipment failures, resulting in longer term limitations, large trading locks and high total overhead expenses.

8.1.3 Horizontal scalability

Scale level means that different software or hardware entities are connected to enhance their capacity to operate together as an integral part of it. Other computers are incorporated into a resource channel and more nodes in the system can be used for horizontal scalability. Service-oriented architecture (SOA) systems and servers are examples of this, as they expand with the addition of more and more servers to the charge system and allow the distribution of incoming applications over all of these systems. Scaling from one server system to another of three systems is one example. In terms of processing power, system architects can design a cluster of tiny computers that outperform computers based on a single conventional CPU. The improved performance of running programmes in the system's extended version is referred to as a flexibility indication.

8.2 SCALABILITY FEATURES

Scalability is a key topic that goes with the ever-expanding modern device infrastructure of the IoT. The IoT network included roughly 24 billion gadgets that can connect to it in 2020. It has an impact on many elements of IoT applications, such as networking and security, identity management, data privacy, big data, vast scalability and so on. Each of these characteristics is a pillar that contributes to the overall scalability anomaly (Kshetri, 2017). Few key topics are covered below.

8.2.1 Business

To make the system more scalable, the database must be able to hold the rising volume of data. A small quantity of data stored in a company later likes to increase its storage capacity in the near future. It should not become outdated and should be able to provide a platform for the increasing amount of data. To adapt and accommodate the future growth, the system technology has to improve. As a result, firms may fail, leaving them with obsolete systems and equipment that must be updated or augmented, which is an expensive proposition.

8.2.2 Marketing

The device should be suited for all situations and, at customer's request, function under all conditions. The device should be simple for the consumer to comprehend and operate, and any device adjustments should be performed promptly and without disturbing the system.

8.2.3 Software

The move between a smaller system and a bigger system and the full benefit of the larger system should be a feature of a scalable system. It should be able, without compromise on service quality, to manage a rising number of connected devices, users, application functionalities or analytics (Jin et al., 2014).

8.2.4 Hardware

Instant remote recognition of gadgets should occur. Masses of comparable equipment should be permitted to utilise explicit customer information compared to different consumers. Gadgets need to connect securely with their backend frameworks, as their data is typically sensitive or secret. Before you leave the device, it is necessary to erase the data. It is important to adequately separate gadgets seized, lost or traded. After vacationing the premises, nobody will pay the expenses of prior tenants.

8.2.5 Networks

In case of deception, a flexible organisation must be able to adjust and operate until the situation is handled. As long as you foresee constant launches and modifications, a complete framework is not necessary for your job. Set up your area and extend your gadget's life. It is a heterogeneous mix that allows PCs and other devices to be connected and has to handle a range of operating systems and protocols. It is also utilised by distant companies using a number of access technologies. For example, a heterogeneous remote organisation is one that can use remote local area network (LAN) methods to manage the cost of support and maintain support while switching to the telephone network.

8.3 TECHNIQUES FOR SCALABILITY

The IoT is a continuously changing subject and flexibility is considered to be a key part of its success. IoT solutions must thus be built to fit numerous connected devices or "things" and to accommodate many different kinds of client and application capabilities. A brief summary of several approaches is provided below (Gao et al., 2018).

8.3.1 Automated bootstrapping

As the number of devices grows, manual functionality such as bootstraps, software and security settings, and device registrations and updates become impractical. Methods based on human contact and facilitation therefore become outdated and ineffectual. In order to save time and operate more effectively, all these services need to automate the aforementioned activities. To first automate the process of booting remotely located devices, the device must have the necessary boot loader, security key and other characteristics.

8.3.2 Controlling the IoT data pipeline

IoT devices create large volumes of data and send them. These data must be logically handled and coordinated to be helpful. A pipeline composed of front-end decisions, a precise placing of information and substantive terminology work is inevitable in IoT applications. These functions are useful for streams of information moving between different frameworks and "objects". These information pipelines need to be designed to address unanticipated floods at the rate of development and execution of information, as the number of gadgets contributing to information creation and interchange is increasing (Dalipi and Yayilgan, 2016). These information stream restrictions should be adjusted to substantial limitations, such as the number of simultaneously connected devices and information streams.

In order to respond to particular support requests, it is essential to monitor the information stream.

8.3.3 Applying the three-axis approach for scaling

X hub-clone scaling, Y hub-various item split scaling and Z pivot-comparable things scaling are the three principal tomahawks for IoT scaling applications. X-scaling refers to the usage of extra assets to spread the incoming demand among various employees, so all employees can process the request. The consolidation of personnel who safeguard state data from request to request is advantageous. It is not difficult to recruit more of these people (Dalipi and Yayilgan, 2016). You have to spread the key works depending on the metric to grow the Y-hub constantly. Z-hub scaling entails assigning responsibility based on future request or response data. As a result, significant part in adaptability due different scaling occurs.

8.3.4 Developing microservices architecture

Small-scale management is a complex configuration process in which big programs are constructed from discrete small cycles that connect with one another using the language rationalism API. It's a good idea to break the program into numerous free trials known as useful units. Each sample has a distinct function. Each of these utilitarian pieces should be able to function on its own. These are features: the units may interact with one another, which is why they are referred to as tiny assistance designs (Saarika et al., 2017).

8.3.5 Adopting multiple data storage technologies

Rather than having a single standard procedure for everything, the IoT framework is made up of numerous components that need distinct functional ways. These applications should be based on the most relevant innovation segments accessible, and each presently employed microadministration should utilise the segment that best meets its needs (Dinh et al., 2018). Furthermore, information questions and recovery requirements should affect the choice of information stockpiling innovation, or, more generally, the choice of dataset.

8.3.6 System expansion

As more devices are added to the framework, it should be simple to scale. Devices must be planned by assessing future needs and creating gadgets to meet them (Dhieb et al., 2020). Allow for long-term use of your organisation's equipment without help (Fang et al., 2012). Support will reduce operational expenses for both devices and businesses in the long run.

8.4 SCALABILITY CHALLENGES AND ISSUES

8.4.1 Protocol and network security

Entry strategies are rules that are used as a standard for establishing plans for transferring data between various PC organisations such as LANs, intranets, and the Internet. As a result, it is obvious that companies must be safeguarded from a wide range of harmful activities. As the number of "things" linked with the Web increases, we must explain new laws that require and remotely identify all "things" (Feng et al., 2018). The importance of convention and organisational security in fostering flexibility cannot be overstated. Then, using the fundamentals of integrating cryptographic calculations, you may obtain high throughput.

8.4.2 Identity management

The executive character is the research field that handles security. In addition to recognising recognised academic members, we also manage access to diverse assets. Recently, we observed a lot of events that perplexed the board's personality by getting access to prohibited data from other sources. Such issues are becoming more common as the number of "things" meant to connect to the Internet rises at an exponential rate. As a result, the general public's acceptability is critical. Another critical factor to consider is approval. If you can't manage who has access to information, everyone will have access to it, which isn't a viable option.

8.4.3 Privacy

Safety contributes to the customer's anonymity and individuality. Individual "things" or personal data must be protected from exposure to the IoT environment, therefore IoT security is a key problem. Because "things" on the IoT freely exchange information, they also interact with other "things". Interoperability is critical in IoT networks because it enables different components of an organisation to function in parallel (Zhang et al., 2019). The sheer number of data that IoT devices may capture is astounding, increasing programmer passages and raising difficult data questions. In reality, manufacturers and programmers may employ devices to physically attack their customers' houses. These issues can undermine the purchaser's personal relevant items and maximise the evolution of the IoT.

8.4.4 Trust and governance

Trust and management are required to identify trust between diverse components and "things", and they are also required from the customer's perspective (Sherly and Somasundareswari, 2015). The IoT must retain trust in the board structure to ensure consumer trust and a foundation for customer trust. From a framework viewpoint, management is critical in determining where the technique should be used and how to handle the control strategy. Even if the board architecture lacks real trust, it might surely permeate across substances or elements. This is presently one of the most important problems under consideration in the study.

8.4.5 Fault tolerance

The primary goal of coordinating for non-critical IoT failures is to respond quickly to changing conditions and to ensure high repeatability. Millions of devices have been delivered, and administrators have been burnt down, putting the IoT susceptible to attack (Talari et al., 2017). Attacks are often carried out by really strong gadgets, and bad frameworks may attempt to obtain power directly or unnaturally against a range of gadgets. This is why, in terms of flexibility, compensating for non-critical failures might be a feasible option.

8.4.6 Access control

Only authorised clients can access various assets inside the IoT framework, thanks to access control. For example, managers have a broader reach than normal clients. As a consequence of the framework's continuing distributed design, new IoT innovations provide less data transmission between IoT tools and the Web, fewer requirements for an IoT tool and one-way access control. You will require the issue. As a result, before they can be incorporated into the IoT, traditional access control techniques must be examined. Later, there are distinct IoT frameworks with distinct architectures.

8.4.7 Big data and knowledge

Big data and IoT can be considered as independent entities. The IoT collects and organises data from an unlimited number of sensors, which it then classifies, details and utilises to make personalised choices. In the future, data collected by IoT devices may offer the most significant threats. Modules are designed to accomplish a specific purpose, rather than being a collection of functionalities that have been merged and compacted into a single device. In some applications, it is possible and preferable to get data using subtraction (Smart Nation Singapore, 2019) instead of using sensors. Because the number of IoT devices is quickly rising, so is the volume of information carried by IoT devices. The importance of data obtained from the customer's operation and behaviour are major obstacle faced by the IoT.

8.5 SCALABLE IOT ARCHITECTURE

- The "Internet of Things" (IoT) is a network of interconnected devices that collects and send data through a distant organisation. Individuals and corporations may take advantage of practically limitless opportunities.

The "thing" may be an associated clinical gadget, a biochip transponder (think animals only), a sunlight-based board, an associated car that warns the driver to a slew of possible support issues or other items that can gather data using sensors and send it to an organisation. This is also an option. Today, IoT has the potential to boost revenue, lower labour costs and improve efficiency are enticing enterprises. Figure 8.1 depicts IoT's adaptive business planning, which is also driven by the requirement for administrative uniformity. Regardless of the explanation (Yang et al., 2019), IoT toolkits give information and knowledge to simplify processes, plan design, robot activities and satisfy coherence needs in a changing business environment, and handle them more effectively. In most IT circles, putting together an adaptable computing structure is a less complicated way to address problems and enable a new layout. Prior to execution, this engineering strategy should integrate the perspectives and ideas of all main partners.

- There are always chances for information to assure the sudden development of customers, business meetings or Internet shoppers. Instead of accepting all requests and paying little attention to traffic, such a framework should remain inactive.
- The IoT framework should also be capable of providing dependable information management while taking into consideration the possibility of diversity. In this case, we must endeavour to give as much information as is reasonably anticipated while maintaining indisputable quality.
- Huge volumes of data may be sent to the rest of the Amazon Web Services (AWS) using any AWS IoT rule. Amazon's online administrations now have a really good layout in the organisation as a result of the substantial quantity of features. AWS thinks that information should be entered from particular data streams, free of mistakes and defects that might hinder full compliance with the framework. The AWS IoT Rules Engine is a fantastic resource for integrating all external endpoints to the IoT core.
- Additionally, because the IoT framework records the flow of information, the AWS IoT Rules Engine is engaged in beginning several activities at the same time. As a result, IoT engineering firms can store several data sets at the same time (Tanwar et al., 2018).
- Always invest in robotic devices that can control several vital equipment at the same time. As the IoT framework expands throughout the organisation, so will the number of intelligent IoT modules. Construction tools, safety precautions and subsequent startup have all grown more automated. As a result, the number of human contacts necessary to eliminate time and money is reduced.
- Architects should also bear in mind that not all information may be required at the IoT treatment phase, when the requirement to separate inputs becomes important. As a result, designers must pay close attention to the sort of information that is in process at the planned stage.

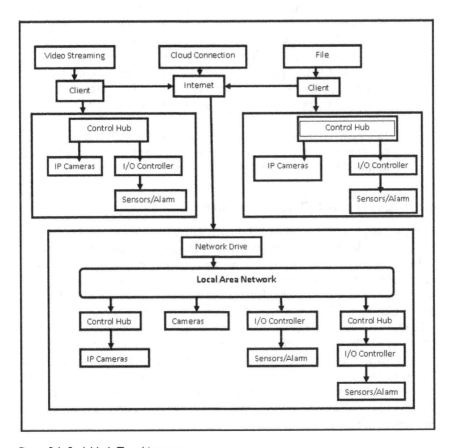

Figure 8.1 Scalable IoT architecture.

- Teams should begin developing each programming section, ensuring that they do not create bottlenecks that negatively impact the possible output. Furthermore, compliance with framework developments should not be an issue.
- Because of the volume of data utilised in engineering design, keeping all of it in a single information repository or archive is not a smart concept. Keep in mind that various types of information indexes may not coexist effectively in the same data stream, therefore search for alternate sources (Talari et al., 2017).
- AWS Green grass tools may be utilised for filtering and capacity plans. Groups may quickly choose unnecessary information that is not required by the framework and delete it by identifying the information instantly in smaller parts. Cloud mixtures are also crucial.

- While this may appear to be a straightforward enough task, there are many moving pieces that must all work together for the IoT to function properly. If the IoT execution is operating effectively, it is important that these numerous communications on the machine operate together.
- Design is a framework that describes the real segments, the necessary organisation's association and layout, the operational processes and the information configurations to be utilised in conjunction with the IoT (Horn, 2016).
- However, because IoT is a choice of advancement, there is no one common reference plan. This means that no one basic chart can be followed for every conceivable execution.

8.5.1 Layers of IoT architecture

Figure 8.2 depicts how different IoT models differ largely in execution; it should be open enough with open norms to assist diverse organisational applications.

The right environment for the IoT has four fundamental components: usefulness, adaptability, accessibility and viability. Versatility is crucial since the framework must be adaptable to the demands of the company or job (Zanella et al., 2014). Despite the lack of universally acknowledged engineering design in the IoT, three-step design is the most fundamental and commonly recognised organisation. It was initially launched after the most recent IoT study was finished. It is divided into three layers: sensing, networking and application (Bakıcı et al., 2013). This is the real engineering layer. This is where the sensors and related devices are developed, which compile various information measures dependent on the project's demands. Sharps, sensors and actuators that communicate with their surroundings are examples of these.

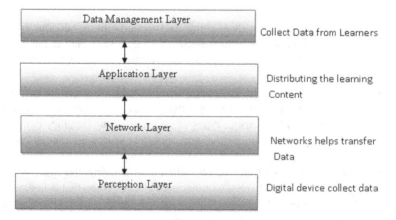

Figure 8.2 Layers architecture.

8.5.2 Network

The data gathered by these technologies should be shared and analysed. This is the organisational layer's duty. Integrate gadgets with other critical aspects, such as personnel and equipment. It is also in charge of all information transfer.

8.5.3 Application

The application layer serves as a conduit between the client and the application. Its role is to offer the client with explicit application administration. For example, there might be a luxury house installation in which consumers tap a smartphone to approve the expresso manufacturer.

8.5.4 Stages of IoT architecture

The following four critical phases are critical for the overall success of IoT implementation.

8.5.5 Connected objects

There would be no IoT if there were no linked or sharp elements. Sensors or remote drives can be used. They react to their surroundings and give data for monitoring (Hernández-Muñoz et al., 2011). Operators go a step further since they can communicate with the environment in a fundamental manner. For example, it may be used to stop valves when water reaches a specific level, or it can be used to remove light as it rises.

Figure 8.3 depicts many steps of the IoT, including gateways, LAN/WAN and sensor gadgets, which must be collected and converted to computerisation once

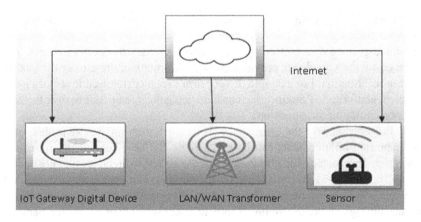

Figure 8.3 Different stages of IoT architecture.

sensors give data (Arasteh et al., 2016). Proceed with processing after obtaining the information. This is the second stage of the IoT design process. The processing of information is nearly complete. The framework for acquiring and changing information determines how information is obtained and modified. The data is sent to sensors and actuators, which are then combined into a sophisticated structure that can be controlled via the portal.

8.5.6 Edge IT systems

The third layer of IoT engineering entails data preparation and improved assessment. The edge IT framework is in charge of carrying out these duties. IoT frameworks necessitate a high quantity of data transfer since they collect a large number of data. As a result, these edge IT frameworks are critical in decreasing the strain on the core IT foundation. Artificial intelligence (AI) and presentation advances are used by edge IT frameworks to offer outcomes from collected data. AI computer gives some knowledge and symbolises innovation in communicating information in an understandable manner.

8.5.7 Data centres and cloud storage

For further internal and external inquiries, data storage is a key element of IoT design. Data should be stored It also enables review and contribution afterwards. Distributed storage is the recommended storage technique for implementing the IoT. More internal processing, which doesn't need to be entered in the cloud or in a physical server farm, can be performed (Gerla et al., 2014). A stronger IT infrastructure can more securely monitor, analyse and store data. This may also be used to integrate sensor data with other information hotspots in order to produce more particular knowledge.

8.6 IOT ARCHITECTURE IN BUSINESS SOFTWARE

Implementing IoT design may greatly enhance the return on investment of bespoke business code. The implementation, on the other hand, will examine more fine-grained data collecting. Information is what drives effectiveness. The more useful the knowledge provided by your information, the greater the quality of your information. For example, if your firm uses a customised inventory management platform, IoT engineering can help you gain greater control over this key component of your organisation.

It also contributes to the expansion of computerisation since more professional programming arrangements can be built, items can be automatically reordered when they approach defined limitations, or use patterns can be thoroughly understood (Hernández-Muñoz et al., 2011). Temperature monitoring applications, for example, might be useful in a pharmaceutical warehouse. It will employ sensors located throughout the distribution centre to assess whether or not the temperature and humidity levels are within acceptable parameters.

This guarantees that critical healthcare supplies fulfil all quality standards before they are delivered. When something goes wrong, an IoT-enabled application can monitor it for you and efficiently deliver a ready alert. Implementing the IoT with a hotel executive structure will be quite beneficial as well. It can collect data from sensors mounted on detergent containers, cappuccino machines, candy machines and other items, allowing employees to check inventory levels in real time.

8.6.1 Basic elements of IoT architecture

Our approach to IoT engineering is reflected in the IoT design diagram. It defines the IoT framework's structural elements and how they relate to information gathering, storage and interaction. The components in the IoT architecture associated with the processor and gate network connected to the cloud portal are depicted in Figure 8.4.

Things: "Things" are things that have sensors. The data generated by these sensors will be shared throughout organisations and actuators, allowing objects to represent themselves. Freezers, streetlights, structures, automobiles, innovative hardware, rescue equipment and everything else are all included in this proposal (Pinna et al., 2018).

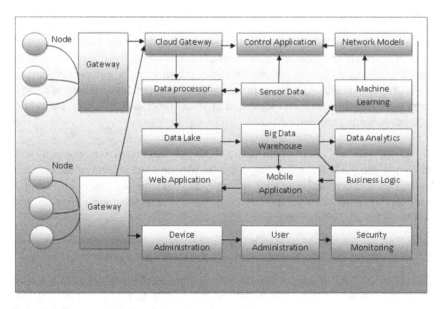

Figure 8.4 Elements in IoT architecture.

Door: Information is transferred to the cloud via the entry channel and vice versa. Routing connects the Internet of Objects things to the cloud components, performs information preprocessing and separation before transferring the information to the cloud, and sends control orders from the cloud to the things. Things then execute orders through their executors.

Cloud: Between the on-site door and the cloud IoT workforce, Entryway employs information pressure and secure data transfer. It also maintains compatibility with other standards and connects with on-site channels that use various conventions depending on whatever conventions the entrance channel supports (Ye-Jin Choi and Lee, 2019). The stream information processor guarantees that the input data is delivered correctly to the information lake and control application. Sometimes data is misplaced or tainted.

The data created by the connecting device is stored in its original format in the information lake. Large quantities of data arrive in "groups" or "streams". When the key experience needs information, it is by no means an information lake, but a major information distribution centre.

Information Distribution Centre: Extract from the information lake the screening and preprocessing information necessary for critical knowledge and send it to the big information distribution centre. A significant information distribution hub was recently cleansed, categorised and coordinated. The information export centre configures data about goods, sensors and things issued by the order control application.

Information survey: Information examiners can utilise data from major distributors to identify trends and gain valuable experience. Following the dissection, a huge quantity of data is displayed, for example, the display of equipment (Zhang et al., 2019), which aids in the detection of faults and the development of an IoT framework. Furthermore, the identified physical links and instances aid in the application's computation.

AI and ML models: Control application models may be created more precisely and efficiently using AI. The modelling is continually updated in light of the chronological information acquired in a massive data warehouse. When information professionals attempt to validate the new model's relevance and efficiency, the new model is utilised to manage the application.

8.7 IOT BUILDING BLOCKS AND ARCHITECTURE

The IoT is a hybrid of progress and technology. Every innovation has its own set of rules within the context of the IoT. Figure 8.5 depicts the position of the relevant network, the basic square sum IoT in the IoT framework and the gadget in this figure. The IoT enables real-world things (IoT gadgets/devices) to interact with

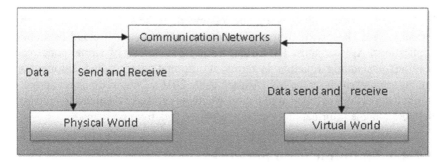

Figure 8.5 Internet of Things illustrated in simple blocks.

the virtual world by implementing business and shared awareness data settings over a communication network (cloud management, stages, and applications). As a result, every IoT architecture incorporates a physical world (Sharma et al., 2017), a virtual world and an organisation.

8.7.1 Things

In terms of the IoT, any article in today's world or data world with new characteristics that may be incorporated in the mail network is deemed something. Things, like virtual things, can become real. Real-world items are commonly referred to as "(IoT) gadgets". The notion of IoT devices differs somewhat from the concept of gadgets. Although objects might be virtual or physical, the term gadget refers to actual hardware that can be supplied to an organisation and is equipped with sensors, actuators, CPU systems, memory and regulators. Things like cloud management plans can be virtual, despite the fact that objects can be virtual (Feng et al., 2018). Virtual things provide programming programs, application programming interfaces (APIs) or application setups for transactions and information about their own location. These applications are a consideration to examine since they have uncommon features or contain attestation keys. Furthermore, virtual entities, such as Amazon Web Services' shadow, can provide enlightening descriptions of actual things. The temperature sensor demonstrates several fundamental concepts.

8.7.2 Physical world

Real and virtual entities can coexist in various parts of the IoT framework. The IoT watch regulators or processors have been the focus of the recent projects or devices. A microcontroller or CPU, a memory limitation, and at least one mail interface are all part of the IoT table. They have information/rendering pins in common and can be linked to at least one sensor, actuator, or mail channel.

8.7.3 Virtual world

The scope of virtual entities within the framework of the IoT is referred to as the virtual world. These virtual items can be obtained via online, cloud or mobile applications, APIs or apps. The virtual world or its variety of virtual goods can play an outstanding function in terms of information recording, information mining, and research on the IoT (Arasteh et al., 2016). Data from the real project is sent to a network or cloud application, where it is erased, revealed and prepared to infer valuable information or acquire required data from participants.

8.7.4 Communication network

In the IoT, the communication network connects the physical and virtual worlds. In essence, the IoT table full of sensors and actuators is not a huge organisation or online network capable of communicating with or allowing network and cloud employees to connect with one another. It's the same as a conventional Web network, and it happens in layers (physical layer, link layer, network layer, transport layer and application layer). To ensure secure and knowledge-rich information transmission, different communication standards are applied to each layer. On a multilevel exchange of information between the real and virtual worlds, the flow of information or any transaction with a specific motive or understanding is shared. The major communication protocols like LR-WPAN, 6LoWPAN, Bluetooth/LE, 802.15.4, LTE, GPRS, CDMA, NFC, Zigbee, 802,11 Wi-Fi and WIRELESSHART are used in the physical layer. Following are the standard application layer protocol that includes DDS, MQTT, REST, CoAP, LLAP, XMPP, SSI, AMQP, XMPP-IOT and MQTT-SN.

8.8 IOT SYSTEM ARCHITECTURE

In unexpected ways, various associations and professional cooperatives design, develop and explain IoT initiatives. In any event, the basic design of the IoT architecture remains unchanged, without regard for implementation or action plans (Miorandi et al., 2012). Engineering fundamentals as illustrated in Figure 8.6, the use of cloud and IoT frameworks is described as a four-layer paradigm.

- IoT devices and gateways
- Communication network
- Cloud or server
- IoT application

8.8.1 IoT devices

As an IoT device, the following requirements are necessary. It is capable of communicating with numerous devices as well as interacting with an organisation

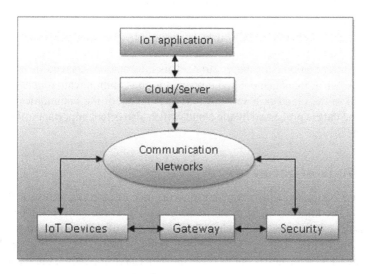

Figure 8.6 Architecture of an IoT system.

online. The gadget should be linked and have firmware or an operating system capable of communicating with other devices. It should be used in conjunction with both measurement and performance. At ambient areas, the sensors can record static or interchangeable data. The measured data or information must be communicated or transferred with a coworker or the cloud (Feng et al., 2018).

The device must include a controller or method to collect the data and archive it as well as a firmware or operating system to store the data or data obtained by the agent or the cloud. To generate standard IoT documents, almost all IoT devices are employed. These might be microcontroller or single core files (single PCs). Arduino, Raspberry Pi, Beagle Bone, CubieBoard, Pinnocio, Banana Pi and other IoT files are among the finest. A microcontroller or CPU, on-board memory (RAM and ROM), high-end and basic GPIO (primarily data-intensive) pins and different communications (such as USB, I2C, SPI, TWI, Ethernet) are all included on the pages. Intelligent microcontrollers or procedures that operate with network interference, RF or mobile phones can also be used to create IoT devices. Some of the main microcontroller firms include Texas Instruments (TI), ARM, Freescale, Intel, Microchip Technology, Atmel and Broadcom.

8.8.2 Sensing and actuating devices

A generic device with integrated processing and communication capabilities inside an IoT application environment. A generic device can process data and communicate with a network via wired or wireless interfaces. These devices primarily collect and analyse data and information from a cloud or server. A generic IoT device might be a Web-controlled industrial machinery or a domestic appliance (Pierro, 2017).

The sensors collect and send data regarding physical quantities such as temperature, humidity, light intensity, strength, density and so on to the on-board processor (Smart Nation Singapore, 2019). This information is saved (temporarily) by the owner or management and delivered to the communication network. It is delivered to the cloud or a server via several layers of the communication network. The data is processed by the cloud, which also offers important information about the drives.

8.8.3 Role of gateways

To set up matching with other devices, the IoT gadget can utilise a door or passless. The importance of doors in the evolution of the convention cannot be overstated. Assume the Zigbee interface is capable of delivering and receiving data via the IoT device. The corresponding devices able to receive and send data using the TCP-IP protocol. In such circumstances (Horn, 2016), the switch will transition from transmitting information via the Zigbee convention to transmitting information via the TCP-IP convention, and the cloud or worker information will be transferred via the TCP-IP convention via the Zigbee convention for IoT device collection. Due to the unique nature of correspondence and organisation on board the IoT gadget, the entrance is like a two-way link between the two organisations.

The transition collects and focuses (sensor) data in accordance with the gadget convention, wraps it in accordance with the correspondence network's convention and pushes the information into the correspondence network for cloud and worker transmission. Furthermore, information, knowledge or data is received and broken down by a cloud or worker, and then measured according to the organisational convention employed in the gadget's organisation and in connection to the cloud information in the IoT gadget. An entryway may be necessary in both cases. These conferences are often conducted at various organisational levels. Zigbee is an actual layer convention, while TCP-IP is a vehicle layer convention, like in the previous model. A network of distant sensors is another example of a device for arranging correspondence across passageways. An IoT device needs to communicate with another IoT device using a different protocol.

Figure 8.7 depicts the corresponding channels of various conventions. The gate offers a circuitous corresponding way between a device and cloud or between a device and another device. IoT endpoints can be discovered together through a door and communicate with different traditional physical layers and connection levels (RF standards such as Bluetooth, Wi-Fi access, Zigbee/ Bluetooth-LE) due to gadget correspondence. The point of entry is referred to as a border crossing.

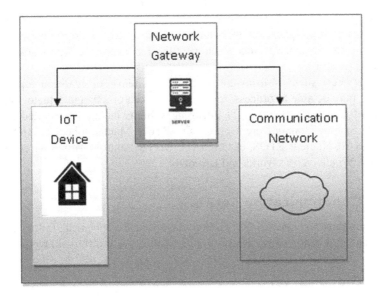

Figure 8.7 IoT Communication through gateway.

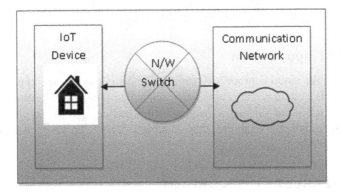

Figure 8.8 Image showing IoT communication without gateway.

8.8.4 Communication gateway

Figure 8.8 depicts the ability of cloud or other IoT devices to be directly connected to IoT devices. In this instance, the device, communication network or communicating devices must all utilise the same protocol to share and exchange data. As a result, no protocol or gateway are necessary. Message queuing telemetry transmission (MQTT), restricted application protocol (CoAP), data distribution service (DDS), AMQP, extensible messaging and presence protocol are examples of application layer protocols that can be used to access the device. Extensible messaging and presence protocols can be utilised in a variety of application layer protocols (XMPP).

IoT devices, such as firmware, operating systems, or real-time operating systems, may be able to process, transmit, store data or manage actuator actions (Gupta et al., 2017). Popular IoT operating systems include Embedded Linux, TinyOS, Snappy Ubuntu Core, Contiki, FreeRTOS, Mantis, mbedOS, RIOT OS, Windows 10, Nucleus RTOS, eCOS, SAFE ROTS, Android Things, Green Hills Integrity, WindRiver VxWorks and BrilloOS.

8.8.5 Communication network

The communication network has different layers such as physical, link, network, traffic and application and protocols of communication at different levels.

8.8.6 Cloud server

The server or cloud is the IoT system's edge. Data from various IoT devices is collected on the cloud and data gathering and analysis are performed to provide relevant insights. He also maintains linked devices and networks, devices utilised for device communication and IoT applications by operating, synchronising and communicating with various IoT devices.

8.8.7 IoT application

The data on the cloud is processed, extracted and analysed by IoT apps. The IoT application is a cloud server program that can collect data, modify data to acquire relevant insights and safeguard insights in IoT devices. An IoT application for home automation, for example, may interpret sensor data and transmit cloud orders to run home equipment.

8.9 SCALABLE AND SECURE IOT

8.9.1 IoT connectivity technologies

Satoshi Nakamoto created a decentralised fake cryptocurrency in a Bitcoin whitepaper in 2008 (Dinh et al., 2018; Fang et al., 2012). Stuart Haber and W. Scotland Stornetta developed the notion of blockchain technology in 1990. Blockchain is a distributed ledger system that enables all network users to exchange transaction records while ensuring dependability without the involvement of a third party (Feng et al., 2018; Gao et al., 2018; Jin et al., 2014). Because all network users can openly view the data on the blockchain, once registered,

Table 8.1 Comparison between the traditional Internet of Things (IoT) and the blockchain-based IoT

Feature	Traditional IoT	Blockchain-Based IoT
Scalability	Limited number of nodes that can be managed and added	No limitations on the number of nodes that can be managed and added
Efficiency	Expensive to process and store data in a central system	Requires data processing and storage costs compared with a central system
Security	Difficult to verify forgery/falsification and restore it	Difficult to forge/falsify

the data cannot be altered or removed, guaranteeing more integrity than previous centralised systems (Kshetri, 2017). Blockchains are classified as public or private based on the methods of participation used by users. The blockchain is open to the public for free, and popular examples include the Bitcoin, Ethereum and EOS blockchains (Dhieb et al., 2020). The public blockchain is now available to the general public. Private blockchains can only enable users who have already approved and allowed participants to join, and the number of nodes is limited and the pace is reasonably quick when compared to public blockchains. Furthermore, private blockchains are classified as blockchain authorisation and blockchain alliance (Pierro, 2017).

The IoT can be installed as a public or private blockchain, depending on the sector and aims. When blockchain technology is used to the IoT environment, it can not only solve cost, scalability and security issues, but it can also achieve better efficiency by assuring the integrity and transparency of the data generated by IoT devices (Saarika et al., 2017; Zhang et al., 2019; Sherly and Somasundareswari, 2015; Smart Nation Singapore, 2019). Table 8.1 depicts some of the benefits of implementing blockchain technology in an IoT context.

8.9.2 Authentication methods of IoT devices

Each device must show its validity and integrity in order to assure the connection safety of devices in the IoT ecosystem without user interaction. Networked device technology makes use of identification (ID) or password (PW)-based authentication, MAC address-based authentication, encryption algorithm-based authentication, challenge–response-based authentication, one-time password-based authentication and certificate-based authentication (Madakam et al., 2008; Miorandi et al., 2012).

8.9.2.1 Identity- or password-based authentication

The most common and basic way to authenticate login names and passwords. The authentication strength of the device is small, the structure is simple, and it is easy to be avoided by attackers.

8.9.2.2 MAC address-based authentication

The media access control (MAC) address uses a unique device ID. Authentication is done by checking the registered MAC address of the device. The security system is quite fast compared to other methods, it is sensitive to data collection and forgery/forgery attacks.

8.9.2.3 Encryption-based authentication

The authentication is accomplished through the use of either a public key algorithm or a secret key algorithm (symmetric key). It performs various functions as a result of the employment of numerous protocols or algorithms. Appropriate procedures should be applied in light of the complexity and time delay. Authentication based on the challenge: the randomly generated challenge value is encrypted. The server sends it as an authentication response value using an algorithm or key. It has a high level of security since it employs random values.

8.9.2.4 One-time password (OTP) based authentication

The device and authentication server are based on the same algorithm and verification authentication is carried out. It is robust against reuse because at authentication it produces a new password and is thus quite safe.

8.9.3 IoT authentication limitations

The conventional IoT device identity verification solution is primarily used to validate and validate the device's validity and integrity. However, the user or the server must intervene, and the device's security level is not taken into account when deciding the device's expansion. Identification or password (ID/PW) and MAC address-based authentication techniques are readily circumvented by attackers, while the following methods gives lower performances based on encryption algorithm, challenge–response and OTP authentication in IoT devices (Pinna et al., 2018). IoT device security verification is time-consuming and expensive and requiring professional or recognised organisation intervention rather than ordinary user intervention will limit its efficiency (Sharma et al., 2017). Furthermore, there is presently no legal and institutional safety certification and assessment system that satisfies the criteria of IoT devices, but in most cases, each product has just a certification and evaluation system. If the IoT device's security is verified via a smart contract, the verification may be depended on without the involvement of a third party.

Therefore, the use of technology is necessary to ensure the safety of known codes and weaknesses in safety and to improve the security of the IoT devices frequently and safely. As a result, in this chapter, we automatically examine and evaluate the security of IoT devices prior to registering them on the block chain, and we offer a technique for safely extending them via automated authentication and device connection.

8.9.4 Scalable and secure IoT

The IoT may be divided into two forms in which the first to deliver user-centric entertainment and later provides services through the installation of convenient programmes like smart telephones or smart TVs. The second is social networks and surroundings such as intelligent grids, intelligent towns and smart houses. Figure 8.8 indicates that IoT devices may be hacked due to numerous faults, but malicious installation apps deployed by users without care or hacker assaults are the most prevalent means of penetration.

Internet-infected devices like Trojan horse infections, rescue goods and roots or Things Internet-infected devices are very vulnerable to network-connected assaults. The next stage of the IoT is the dispersed self-organising networking of a single device and its neighbours. Sure and scalable IoT connection solutions are necessary for this context. IoT devices must have trust in the automated certification and growth of IoT devices in the security level and security evaluation systems.

Figure 8.9 shows the architecture of IoT security level for evaluating the security of IoT devices, which is described in the following steps.

Step 1: The IoT device manufacturer checks the software that will be installed on the IoT device ahead of time and creates a whitelist. A whitelist that

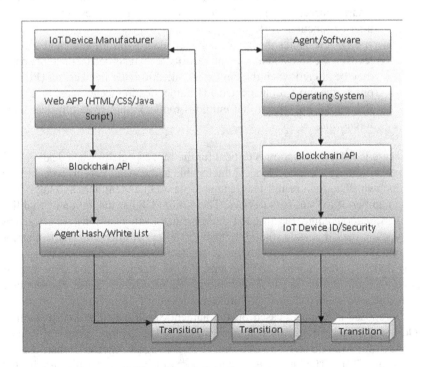

Figure 8.9 Architecture of IoT security level.

meets the requirements of each IoT environment can be created for the same IoT type. The manufacturer's whitelist and the initial hash value of the agent (IAHV) provided in the IoT device form the basis of the intelligent contract. Manufacturers of IoT devices will access intelligent contracts via decentralised apps such as web pages (dApps). Internally, blockchain may simply be accessed via an API.

Step 2: The manufacturer uses the whitelist smart contract to register the whitelist and IAHV of the agent deployed in the IoT device on the blockchain (WSC). IoT devices may utilise WSC to query the blockchain information to determine if the agent's IAHV matches the device agent hash value of the installed agent (DAHV). By comparing it to the proxy hash value stored on the blockchain, it assures that the proxy has not been tampered with.

Step 3: Manufacturers are compensated with tokens for creating and updating the WSC.

Step 4: The agent of the IoT device verifies whether unverified apps are loaded using the device's agent information and the whitelist posted on the blockchain through the WSC. The smart contract is nearly complete, so IoT devices will receive the most recent whitelist.

Step 5: The agent transmits the security state of the IoT device evaluated by it, together with the device's unique identifying information, to the scoring smart contract (SSC) and the device's security level is established via an internal function.

Step 6: The SSC stores the device's unique identity information and security level on the blockchain, and the security level recorded on the blockchain may be altered when the device is linked to other devices. on (PUF) is one of the most sophisticated IoT authentication techniques. PUF is an IoT digital fingerprint that employs random nano-scale physical interference. E

Secure identity verification is critical for the IoT, especially when dealing with a large number of devices. Physical unclonable authenticati very irrevocable IoT has its own PUF. As a result, PUF can provide unique identifiers and keys for a large number of IoT network devices. Therefore, PUF has the option of enabling IoT certification online. Because IoT devices run on validated software created by the vendor, their first state study is a "pure" one free of infection. Authentication technology allows "pure" IoT devices to connect securely to other devices.

If malware contaminates IoT devices, the "contaminated" IoT devices may have an influence on the IoT adjacent in the network, because conventional authentication cannot identify pollution and prevent it from spreading to others. The proposed solution is based on the authentication stated earlier, as well as additional criteria for secure communications and scalability. That is the smart contract for whitelisting and scoring device integrity checks (Smart Nation Singapore, 2019).The suggested security assessment method validates the IoT device's security twice: first by validating the integrity of the agent deployed in the device, and then by passing the validated agent again.

8.10 SCALABILITY PERFORMANCE ON A LARGE-SCALE NETWORK

8.10.1 Performance evaluation

To establish if a device is legitimate, most conventional IoT device authentication techniques will check for the existence of a key. However, when the network grows, this identity verification technique does not solve the security issue and requires manual participation. As a result, we developed a trustworthy security assessment system that combines whitelists and smart contracts to automatically check security and then registers it on the blockchain. Because of the diverse uses of IoT devices, security and scalability needs differ. As a result, when security is high, the suggested paradigm provides maximum scalability and restricts scalability when security is weak.

8.10.2 Evaluation method

An abstract model is presented to evaluate the performance of the recommended approach. Two hundred IoT device nodes are randomly created and put on 30×30 two-dimensional planes for network simulation. Following the configuration of each IoT device's location information, these devices can only connect with other devices that can communicate physically. The agent and smart contract are expected to record the security score on the block chain. Security scores have been issued to all 200 nodes, and their security levels are A, B, C and D. In addition, four malware-infected malicious nodes (two Cs and two Ds) were put up. Furthermore, in order to assess security, the integrity of all software will be validated against A-level nodes, where the installed software is correctly verified, so that even if it is linked to potentially hazardous devices, it will not be infected with viruses.

Even a basic connection might be infected with viruses if it installs unapproved software. Furthermore, untrusted malicious software will have a major influence on the security of IoT devices in use. As a result, when 200 nodes make a single connection inside this range, the security score must drop from 10 to 19, with a chance of 1/50, based on the assumption of unchecked software, that is, new software that has not emerged.

The suggested approach calculates the security score and level depending on the amount of unconfirmed applications. The lower the level, the more limitations there are on expansion. Table 8.2 provides the safety values as well as the degrees of unverified software. The proposed model includes discrete connection areas depending on level and can only connect to devices of the same level within the connection range. The comparative approach (i.e. usual model), on the other

Table 8.2 Different stages of connection range

Range	Security level	Connection level existing model	Connection level proposed model
100-81 (L0)	A	4 hop	4 hop
80-61 (L1)	B	4 hop	3 hop
60-41 (L2)	C	4 hop	2 hop
40-21 (L3)	D	4 hop	1 hop
20-0 (L4)	E	4 hop	0 hop

Table 8.3 Different level of whitelist update period

Score Level	Period		
	0.5	1.0	1.5
Level A	5	10	15
Level B	10	20	30
Level C	15	30	45
Level D	20	40	60
Level E	25	50	75

hand, tries to expand IoT devices to all connected devices while disregarding device security (Bakıcı et al., 2013). The proposed model checks to verify if new software has been installed and if the WSC agent has given the security level. It will be extended in accordance with the level of security. Request an update from the whitelist if there are other IoT devices within the connection range but the connection is not secure enough.

If device connection failures persist within the connection range, the whitelist will be updated to check the software, and the security score will be randomly increased from 10 to 19 points. When the whitelist is updated to remove malware, the security score rises to improve scalability. Despite the whitelist and other flaws, the program has not been tested, but one vulnerability has been fixed (Zhang et al., 2019). There are several software scenarios.

The suggested model's scalability and security outcomes may vary depending on the whitelist and update cycle; to get an appropriate value, the greatest value of network throughput is selected as the basic value of the update cycle (Talari et al., 2017). The throughput simulation results based on the update cycle are depicted in Figure 8.10. As indicated in Equation (8.1), the throughput is computed by dividing the successfully delivered data size by the data transmission time:

$$\text{Throughput} = \text{data size successfully sent} / \text{data transmission time} \qquad (8.1)$$

As transmission latency increases, so does the overhead required to transfer data, while throughput decreases. In fact, when overhead is decreased, throughput

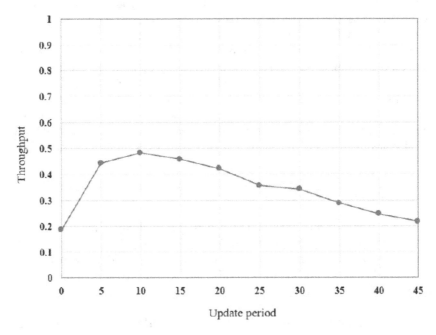

Figure 8.10 Throughput simulation.

increases. IoT devices with high scalability can transport data to their destination quickly and efficiently, resulting in higher throughput. Because of the limited number of connections, throughput will be decreased. The simulation findings demonstrate that raising the update duration reduces the data packet overhead needed for the update, boosting network throughput. However, only by increasing the update cycle to a particular degree can throughput be enhanced (Horn, 2016). When the update time is too long, the network's scalability is constrained, resulting in a reduction in throughput. As a result of the simulation results, the basic update cycle is set to 10 times, because the maximum throughput is attained when 10 connection failures accumulate, and then the whitelist is updated. Because the given update cycle varies depending on the usage and requirements of IoT devices, we examined the basic cycle 0.5 and 1.5 times in addition to comparing models in this study. Table 8.3 shows the criteria for raising the scores of each level based on the whitelist's update time. The proposed model splits the value of the evaluation update period into 5, 10 and 15, which are 0.5 times and 1.5 times the basic period and the basic period, respectively.

8.10.3 Evaluation results and analysis

In this study, each node is allowed to automatically connect to devices within the connection range in order to test the scalability of the suggested model and the comparison model (Yang et al., 2019). The current model allows any devices within the connection range to connect to each other; however, the proposed model only allows them to connect to the same level of equipment within the connection range based on the security level provided by the authorised technology. The conventional (comparison) model is represented by C, while the recommended models with update cycles of 5, 10 and 15 are represented by P1, P2 and P3, respectively. Figure 8.7 depicts the outcome of comparing the proposed model's security to the security of the comparison model when 200 IoT devices in ad-hoc mode attempt to connect to devices within the connection range. The vulnerability is calculated by dividing the number of affected devices by the total number of devices in Equation (8.2):

$$\text{Vulnerability} = \text{Malicious devices/All devices} \qquad (8.2)$$

The findings shown in Figure 8.11 compare the scalability of the current technique to the previous paradigm when 200 devices attempt to connect to devices within the connection range. Scalability is measured by dividing the number of connected IoT by the number of physically available connections, as indicated in Equation (8.3):

$$Scalability = Connected\ IoTs/Physically\ available\ connections \qquad (8.3)$$

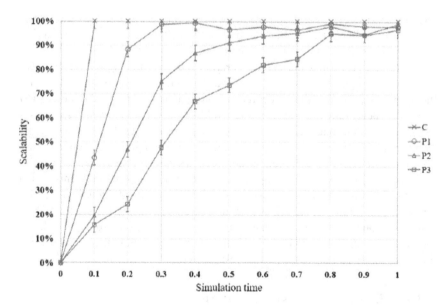

Figure 8.11 Throughput scalability.

The number of runs is shown by the X-axis, while the scalability of the linked equipment is represented by the Y-axis (Zanella et al., 2016). The standard deviation is shown by the error bar. When linked to all devices within the connection range, their value is 100 percent. Because devices can only be joined to devices of the same level, and the connection level is limited, the proposed technique has relatively poor scalability in the early connection stage when compared to the analogous concept. When devices with low security levels seek to connect to other devices, they are limited to connections, necessitating the whitelist to be frequently updated. The longer the period between whitelist updates, the longer it will take for devices to expand by more than 90 percent. As demonstrated in Figure 8.8, P1, 0.5 and P3 each require 0.23 and P2 0.8 before scalability reaches 90 percent or greater.

8.11 IOT APPLICATIONS AND TRANSFORMATIONS

8.11.1 IoT applications

This section will cover several IoT applications. Many apps and services can and are being utilised in the IoT (Miorandi et al., 2012). We'll go through a few of them here. The works referenced offer extensive descriptions. Before digging into IoT security risks and attacks, as well as security requirements, it is crucial to remember that the purpose of this section is to demonstrate instances of IoT application areas where different hazards and assaults may occur.

8.11.2 Health care systems

Health care systems are becoming more reliant on smart apps that allow the IoT (Sharma et al., 2017) as WSN, RFID, smart wearable devices and sensors (such as FitBit [Meng et al., 2018]) gain popularity. Patient monitoring and medication administration may be automatically adjusted and maintained in this smart health care system without the need for direct physical intervention. The health care system was previously buried behind a secure network design. They now work in a more open environment as a result of the IoT (Feng et al., 2018). Using wearable blood pressure monitoring devices, the patient's blood pressure and other data may be uploaded to the hospital database and analysed on a regular basis by appropriate physicians. It can then be used to make a diagnosis and decide on a treatment plan. For example, by using "biobands," wearable wristbands and shoe clips to monitor heart rate, the user's medical data, such as heart rate, blood oxygen saturation or sleep analysis, may be reliably captured and preserved. A smartphone application may be used to operate and monitor this gadget.

8.11.3 Smart homes and buildings

Smart homes employ home devices that enable the IoT to offer a more flexible and comfortable living environment (Zhang et al., 2019). Smart sensors, for example, may utilise physiological measures to predict a person's emotional state and adjust the room atmosphere accordingly. The sophisticated electric heater can automatically regulate the temperature of the room without the need for personal involvement. The smart metre may automatically communicate measurement data to the billing company. There are a number of real-world applications available, such as the "CURB" (Kshetri, 2017) energy intelligence system, which allows users to remotely adjust the temperature in their houses. It can also determine which devices are turned on and how much power they consume over a certain time period. Based on this data, it can anticipate future utility expenses. Another example is the "Philips Hue" (Meng et al., 2018) wireless lighting system, which allows users to control lights, adjust brightness, create timers, build routines and even change colours via mobile applications.

8.11.4 Smart transportation

Intelligent transportation systems aid in the monitoring and adjustment of traffic data (between cars and transportation infrastructure), the calculation and integration of this data in real time, the interaction with the transportation network for analysis and evaluation, and the adjustment or assistance of the vehicle itself. GPS and RFID tracking systems are extensively used (Saarika et al., 2017). "B-Scada" (Zhang et al., 2019), for example, is a system-wide data management architecture for intelligent transportation systems that supports the IoT. It takes real-time data from many sources, analyses it and implements solutions such as transportation route realignment, among others. In terms of efficiency and cost, the IoT may also enhance scheduling, freight distribution and fuel utilisation.

8.11.5 Smart grid

The smart grid is a type of smart infrastructure that aids in electricity distribution, management and consumption. It includes a wide range of energy and operation metrics, such as smart metres, smart devices and different energy efficiency (Feng et al., 2018). From the central core to the edge network, smart grid technologies comprise smart systems for power distribution and control. This will aid in meeting energy-saving goals by utilising low-cost, low-power IoT devices. Some initiatives (Ahamad et al., 2019) aim to minimise carbon emissions while also achieving great energy efficiency (Jin et al., 2014).

8.11.6 Smart city

A smart city is a pervasive system that includes various applications that support the IoT as well as services relating to health, construction, transportation and public utilities. These apps and services work together to benefit the metropolitan area (Arasteh et al., 2016). The goal is to establish an environment (through the use of information and communication technology) that will improve the quality of life for people living and working in cities, as well as the interaction between various entities, systems and applications (Dhieb et al., 2020). Simultaneously, it will aid in the management of the economy, the environment, liquidity and municipal infrastructure and services. Several projects have been launched to develop smart cities that are compatible with the IoT. Some examples are Singapore Smart Nation (Feng et al., 2018), Amsterdam Smart City (Gao et al., 2018) and Barcelona Smart City (Tanwar et al., 2018). These initiatives have resulted in a real-world smart city experience through sustainable space development, smart digital linkages and enhanced IoT services.

8.12 THREATS AND ATTACKS

The potential hazards and threats associated with IoT, as well as the myriad application possibilities are already covered (in Section 8.4). Several papers (Pinna et al., 2018) have looked into the security of the IoT, as well as weaknesses and threats (Madakam et al., 2008; Kshetri, 2017) have been classified as potential risks and attacks based on the IoT architecture's multiple tiers. Some of them (Gerla et al., 2014; Meng et al., 2018) separate risks and attacks from specific security problems like as identity, access control, trust, middleware and mobility. Some of them (Pierro, 2017; Saarika et al., 2017) additionally categorise risks and assaults according to their application and use case circumstances. Furthermore, (Sharma et al., 2017) categorises numerous security flaws in IoT systems based on the kind of IoT architecture, such as centralised, collaborative, connected and decentralised IoT.

We argue, however, that the classification is vague and fails to account for the differences between the many attack scenarios that occur in IoT and traditional distributed systems. The IoT poses new obstacles to the development of secure and reliable solutions. Because of the size and complexity of IoT systems, it is difficult to build a threat and attack model using commonly used security methodologies. According to the findings, it is challenging to adapt current threat modelling approaches, such as those used in ordinary computer systems, to the IoT. We next go over the essential components of the IoT ecosystem in detail, categorising security risks and attacks that fall within these categories.

8.12.1 Category

Attacks on an IoT system can target a wide range of vulnerabilities (Saarika et al., 2017; Zhang et al., 2019; Sherly and Somasundareswari, 2015; Smart Nation Singapore, 2019; Talari et al., 2017). These include everything from the devices themselves to the communication between them and the services and apps offered. Users, as well as the fundamentally mobile and dynamic character of

Table 8.4 Category and its threats and attacks

Category	Threats and attacks
Communications	Wired or wireless mediums such as routing channels and data transmission
Device/services	Physical IoT devices and associated low-level services
Users	Human being in an IoT system
Mobility	Different network domains
Resources	Heterogeneous infrastructure

these systems, are at risk. When developing security needs and an appropriate security architecture for IoT systems, it is necessary to analyse the many possible risks and assaults on these systems in order to discover an acceptable set of criteria and the resultant security architecture. Based on the features of the IoT, we have classified potential risks and assaults into five categories: communication, devices / services (Miorandi et al., 2012), users, mobility and resource integration.

These classifications are shown in Table 8.4, which displays the classifications of devices and attackers. We use the word "services" in its broadest meaning, whereas "resource integration" refers to applications that rely on a variety of devices and services to satisfy the demands of end users. The IoT, according to this notion, is made up of communicating users and devices, with the devices organising services. To fulfil the demands of end users, devices and their services are constructed (service integration / splitting / composition). Devices and users can both be mobile (and dynamic). Threats and assaults on wired and wireless media services, as well as physical IoT devices and the related low-level services, are included in the communication category. Users protect themselves against threats and assaults on IoT users. The term "mobility" refers to threats and attacks that take advantage of the movement of IoT devices and smart gadgets. Finally, resource integration looks into the dangers and attacks that might occur as a result of integrating diverse services into end-user applications.

Figure 8.12 shows the logical and the technological components of an IoT system. It is important to understand that a "communications" attack can alter a packet to implant malicious code that takes control of a device. The example represents an attack on two different parts of an IoT system in communication and devices. Numerous other similar attacks that involved several aspects. These categories are discussed in depth.

8.12.2 Communications

Communication is the heart of the IoT, which links between users and devices. The most common threats to IoT are router assaults, active data attacks, passive data attacks and floods. The attackers target routing protocols and network traffic in order to disrupt data flow or divert the routing path to an unsecured target

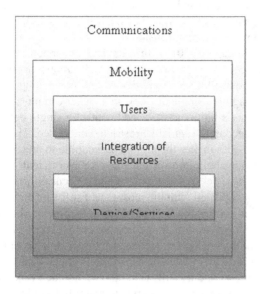

Figure 8.12 Threats and attacks category.

in router. These attacks are frequently carried out via black holes, wormholes and poisons (Tanwar et al., 2018). The black hole attack infects the network by prohibiting all network connections.

An attacker node grabs data packets at one point in the network and tunnels them to one or more targets without the sender's knowledge. The attacker node employs it in an Internet phishing assault to collect sensitive and private information from a victim. Malicious code is frequently injected into the victim's computer in order to breach the computer system. Instead of weakening network routing, active data assaults alter or remove data by directly targeting genuine data packets. Channel jamming and various forms of data manipulation are examples of these attacks, which may or may not result in valid packets. Active data attacks can target the payload, the header, or both of a packet. Passive data attacks look for information without affecting the content of the communication. Wiretapping and traffic analysis are two examples (Gerla et al., 2014). In the event of an eavesdropping attack, the attacker node passively listens to information that is transmitted through the network between the sender and recipient. The attacker node intercepts network data and examines it to determine location, message frequency, underlying payload traffic and host identity.

8.12.3 Device/services

The common threats to IoT system equipment services are physical assaults, device subversion attacks, device data access and device degradation. The great majority of IoT devices are used in public, and security threats such as equipment damage and disconnection are common. An attacker might physically disconnect an IoT device from the Internet, causing irreversible damage or even destroying it (Smart Nation Singapore, 2019). In a device subversion attack, the attacker takes complete or partial control of the device. There are two types of controlling IoT devices: controlling a single device and controlling multiple devices. In the former (managing a single device), an attacker can physically or electronically access the user's home network and control a single device. As a result, its functions may become unavailable, restricted or abused. The low power consumption of IoT devices makes them more susceptible to attacks, partly because the security protections built into such devices are insufficient.

Furthermore, if these devices contain embedded security features, they are frequently unable to be updated with the most recent software and security updates. We believe that this sort of attack is common among networked computers and is not limited to IoT devices. The limited nature of IoT devices makes them more vulnerable to assaults, which is exacerbated by poor protective mechanisms for these devices. It is difficult to upgrade applications or patch new security features to deal with new threats and assaults using this technique (Yang et al., 2019). An attacker in the latter instance can manipulate many IoT devices and modify services, such as many linked IoT devices (things controlled by humans). An attacker may disrupt the traffic monitoring service by manipulating a number of underlying sensors (Ye-Jin Choi and Lee, 2019). Device downgrade is a DoS attack that aims to limit access to a service by targeting the device's operation rather than the network's capacity to handle traffic. In a typical DoS attack, the service is overwhelmed by dealing with bogus traffic, yet a single node is unaffected. This issue, however, is more crucial in the case of the IoT. Therefore, large-scale device downgrade attacks on these resource-constrained IoT devices may cause resource unavailability and bring the entire system to a halt (Feng et al., 2018).

8.13 CONCLUSION

The Internet of Things (IoT) is a fast-evolving technology that provides a platform for end users to build a variety of cost-effective, efficient and user-friendly applications and services. Security, on the other hand, is one of the most important hurdles to broad adoption of IoT devices. In this work, we examined and assessed current IoT security requirements, and the provided a set of security architecture criteria for IoT. We have clearly highlighted the issues that must be addressed while designing such a secure IoT architecture in accordance with the criteria provided. The existing IoT security standards are incapable of dealing with a large number of threats and assaults in an orderly manner. A systematic approach is needed to address IoT security requirements that can capture the basic standards for building a secure IoT architecture.

Unlike prior IoT security evaluations, the various security issues have been classified into five distinct categories such as threats and assaults on communications, equipment/services, users, mobility and resource integration, and their potential of each area has been evaluated. These categories and their associated solutions, as well as a list of important security criteria for securing IoT systems, have been produced. Many touted benefits, such as scalability, availability, reliability, trust, identity and so on, are easier to accomplish. It should be highlighted that if the IoT system is not adequately assessed for possible threats and does not contain adequate security requirements. The security issues that exist in each layer of the IoT security architecture are studied. The security requirements are present in the IoT security architecture's five distinct layers, which include the sensor layer, network management layer, service composition layer, application layer and user interface layer.

Bibliography

Ahmad, M., Younis, T., Habib, M.A., Ashraf, R. and Ahmed, S.H. (2019) Scalability in Internet of Things: Features, Techniques and Research Challenge. Scalability in Internet of Things: Features, Techniques and Research Challenge: A Review of Current Security Issues in Internet of Things. In *Recent Trends and Advances in Wireless and IoT-enabled Networks*. Berlin/Heidelberg: Springer, pp. 11–23.

Arasteh, H., Hosseinnezhad, V., Loia, V., Tommasetti, A., Troisi, O., Shafie-khah, M. and Siano, P. (2016) IoT-based Smart Cities: A Survey. In Proceedings of the IEEE 16th International Conference on Environment and Electrical Engineering (EEEIC), Florence, Italy, 7–10 June 2016, pp. 1–6.

Bakıcı, T., Almirall, E. and Wareham, J. (2013) A Smart City Initiative: The Case of Barcelona. *J. Knowl. Econ.*, 4, 135–148.

Briante, O., Cicirelli, F., Guerrieri, A., Iera, A., Mercuri, A., Ruggeri, G., Spezzano, G. and Vinci, A. (2019) A Social and Pervasive IoT Platform for Developing Smart Environments. In *The Internet of Things for Smart Urban Ecosystems*. Cham, Switzerland: Springer International Publishing, pp. 1–23.

Chayan Sarkar, S.N. Akshay Uttama Nambi, R. Venkatesha Prasad and Abdur Rahimy (2017) A Scalable Distributed Architecture Towards Unifying IoT Applications. Scalability in Internet of Things: Features,Techniques and Research Challenge from Delft University of Technology. *International Journal of Computational Intelligence Research.*13(7), 1617–1627.

Christidis, K. and Devetsikiotis, M. (2016) Blockchains and Smart Contracts for the Internet of Things. *IEEE Access*, 4, 2292–2303.

Dalipi, F. and Yayilgan, S.Y. (2016) Security and Privacy Considerations for IoT Application on Smart Grids: Survey and Research Challenges. In Proceedings of the IEEE 4th International Conference on Future Internet of Things and Cloud Workshops (FiCloudW), Vienna, Austria, 22–24 August, pp. 63–68.

Dhieb, Najmeddine, Hakim Ghazzai, Hichem Besbes and Yehia Massoud (2020) IEEE Technology and Engineering Management Conference (TEMSCON20), Detroit, USA.

Dinh, T.T.A., Liu, R., Zhang, M., Chen, G., Ooi, B.C. and Wang, J. (2018) Untangling Blockchain: A Data Processing View of Blockchain Systems. *IEEE Trans. Knowl. Data Eng.*, 30, 1366–1385.

Fang, X., Misra, S., Xue, G. and Yang, D. (2012) Smart Grid – The New and Improved Power Grid: A Survey. *IEEE Commun. Surv. Tutor.*, 14, 944–980.

Feng, L., Zhang, H., Chen, Y. and Lou, L. (2018) Scalable Dynamic Multi-Agent Practical Byzantine Fault-Tolerant Consensus in Permissioned Blockchain. *Appl. Sci.* 8(10), 1919; https://doi.org/10.3390/app8101919

Gao, F., Zhu, L., Shen, M., Sharif, K., Wan, Z. and Ren, K. (2018) A Blockchain-Based Privacy-Preserving Payment Mechanism for Vehicle-to-Grid Networks. *IEEE Netw.* 32, 184–192.

Gerla, M., Lee, E., Pau, G. and Lee, U. (2014) Internet of Vehicles: From Intelligent Grid to Autonomous Cars and Vehicular Clouds. In Proceedings of the IEEE World Forum on Internet of Things (WF-IoT), Seoul, Korea, 6–8 March 2014, pp. 241–246.

Gupta, Anisha, Christie, R. and Manjula, R. (2017) Scalability in Internet of Things: Features. Techniques and Research Challenges. *International Journal of Computational Intelligence Research* 13(7), 1617–1627 ISSN 0973-1873.

Hernández-Muñoz, J., Vercher, J., Muñoz, L., Galache, J., Presser, M., Hernández Gómez, L. and Pettersson, J. (2011) Smart Cities at the Forefront of the Future Internet. In *The Future Internet*. Domingue, J., Galis, A., Gavras, A., Zahariadis, T., Lambert, D., Cleary, F., Daras, P., Krco, S., Müller, H., Li, M.S., et al., Eds., Lecture Notes in Computer Science. Berlin/Heidelberg: Springer. Volume 6656, pp. 447–462.

Horn, John (2016) 3 Keys to Ensuring Scalability in the IoT www.itproportal.com/2016/02/02/3-keys-to-ensuring-scalability-in-the-iot

Jin, J., Gubbi, J. and Marusic, S. (2014) An Information Framework for Creating a Smart City Through Internet of Things. *IEEE Internet Things J.* 1, 112–121.

Krylovskiy, A., Jahn, M. and Patti, E. (2015) Designing a Smart City Internet of Things Platform with Microservice Architecture by Future Internet of Things and Cloud (FiCloud), 3rd International Conference.

Kshetri, N. (2017) Can Blockchain Strengthen the Internet of Things? *IT Prof.*, 19, 68–72.

Madakam, S., Ramaswamy, R. and Tripathi, S. and IT Applications Group, Internet of Things (IoT) (2008) A Literature Review, National Institute of Industrial Engineering (NITIE), Vihar Lake, Mumbai, India http://file.scirp.org/pdf/JCC_2015052516013923.pdf

Meng, W., Tischhauser, E.W., Wang, Q., Wang, Y. and Han, J. (2018) When Intrusion Detection Meets Blockchain Technology: A Review. *IEEE Access* 6, 10179–10188.

Miorandi, Daniele, Sabrina Sicari, Francesco De Pellegrini and Imrich Chlamtac (2012) CREATENET, via Alla Cascata 56/D, IT-381, 123 Povo 207, Internet of things: Vision, applications and research challenges b, Trento, Italy b Dipartimento di Informatica e-Comunicazione, Università degli Studi dell' Insubria, via Mazzini, 5, IT-21100 Varese, Italy.

Pierro, M.D. (2017) What Is the Blockchain? *Comput. Sci. Eng.*, 19, 92–95.

Pinna, A., Tonelli, R., Orrú, M. and Marchesi, M.A. (2018) Petri Nets Model for Blockchain Analysis. *Comput. J.* 61, 1374–1388.

Saarika, P.S., Sandhya, K. and Sudha, T. (2017) Smart Transportation System Using IoT. In Proceedings of the 2017 International Conference on Smart Technologies for Smart Nation (SmartTechCon), Bengaluru, India, 17–19, pp. 1104–1107.

Sharma, Pradip Kumar, Saurabh Singh, Young-Sik Jeong and Jong Hyuk Park (2017) DistBlockNet: A Distributed Blockchains-Based Secure SDN Architecture for IoT Networks, IEEE Communications Magazine 55, 78–85.

Sherly, J. and Somasundareswari, D. (2015) Internet of Things Based Smart Transportation Systems. *Int. Res. J. Eng. Technol.*, 2, 1207–1210.

Smart Nation Singapore. (2019). Available online: www.smartnation.sg (accessed on 12 October 2019).

Talari, S., Shafie-khah, M., Siano, P., Loia, V, Tommasetti, A. and Catalão, J.P.S. (2017) A Review of Smart Cities Based on the Internet of Things Concept. *Energies* 10, 421.

Tanwar, S., Tyagi, S., Kumar, S, Hu, Y.C., Tiwari, S., Mishra, K.K.,Trivedi, M.C., Eds. (2018) The Role of Internet of Things and Smart Grid for the Development of a Smart City. In *Intelligent Communication and Computational Technologies.* Springer: Singapore, pp. 23–33.

Yang, Q., Yang, T. and Li, W., Eds. (2019) Internet of Things Application in Smart Grid: A Brief Overview of Challenges, Opportunities, and Future Trends. In *Smart Power Distribution Systems.* London: Academic Press, pp. 267–283.

Ye-Jin Choi, Hee-Jung Kang and Il-Gu Lee (2019) Scalable and Secure Internet of Things Connectivity, *Electronics*, 8, 752, doi:10.3390/electronics8070752

Zanella, A., Bui, N., Castellani, A., Vangelista, L. and Zorzi, M. (2014) Internet of Things for Smart Cities. *IEEE Internet Things J.*, 1, 22–32.

Zhang, Daqiang, Laurence T. Yang and Hongyu Huang (2019) Searching in Internet of Things: Vision and Challenges. ieeexplore.ieee.org/abstract/document/5951906/ Scalability in Internet of Things: Features, Techniques and Research Challenges

Chapter 9

Applying fuzzy logics to detect reliable sensors on IoT

M.S. Nidhya, M. Prasad, M.D. Javeed Ahamed, and Shaik Bajidvali

CONTENTS

9.1 INTRODUCTION

IoT is composed of a number of sensors. Generally, sensors are scattered in an environment. For example, the data about the environment is collected and the information sent to the sink node or a base station. In IoT, there are hundreds of sensors used to monitor various locations, buildings, traffic, hospital, agriculture and so on. The application of sensors in IoT at times is not reliable.

Sensors are very tiny and have very small processors. Energy and time delay are major factors for the sensors to work efficiently in an environment. Generally, there are three types of architecture followed for sensor networks.

1. Hierarchical or clustered architecture
2. Flat architecture
3. Location based

9.1.1 Clustered architecture

Clustered or hierarchical network is a network in which nodes are grouped as a cluster and each cluster has a cluster head and a number of nodes.

DOI: 10.1201/9781003335801-9

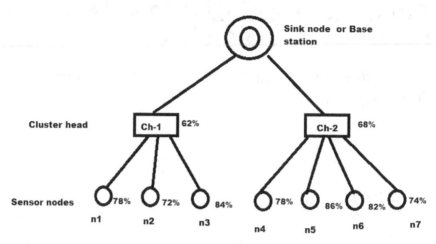

Figure 9.1 Clustered network or hierarchical network.

In Figure 9.1, it is clearly illustrated that a network has one base station or sink node, and two cluster heads, Ch-1 and Ch-2, and there are seven nodes. These seven nodes are grouped into two clusters. Nodes n1, n2 and n3 are under the control of Ch-1 and nodes n4, n5, n6 and n7 are controlled by Ch-2. Sensor nodes inside the clusters are monitoring the environment and collect the sensed data from the environment and send the data to the cluster head. Then the cluster head will pass the data to the base station. From the base station or sink node, the end user can collect the data.

In a clustered network, nodes cannot send the collected data directly to the sink node. They can communicate only through the cluster head. Cluster head is selected based on the energy conservation of a node. A node can be elected as a cluster head based on the energy it retains. The main constraint for a cluster head is energy.

In a clustered architecture, the node that loses its energy quickly is the cluster head when compared to other nodes. Other sensor nodes are monitoring the environment and collect the data from the environment. Then they forward it to the cluster head. The cluster head node aggregates the information from the sensor node and passes the information to the sink node. The sink node may send a query about the environment or about the sensors. Those queries are also solved and answered by the cluster head. The cluster head is always the active one when compared with other nodes in the network.

In Figure 9.2, the energy of each node and that of cluster head is clearly illustrated. When compared to other nodes, the cluster head has a very low energy of 62 and 68 percent. If cluster head's energy drops below 60 percent, then the cluster head will be changed and it is elected within the cluster. The node which has a high energy in a cluster is elected as the cluster head. In the first cluster

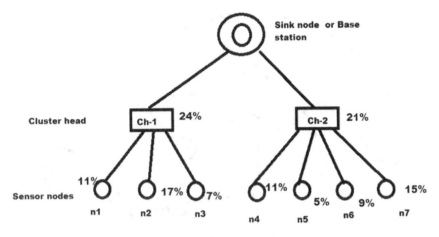

Figure 9.2 Energy value of sensor node and cluster head.

Ch-1, node n3 has a high energy when compared to other nodes. So that node is elected as the cluster head.

In a second cluster Ch-2, the energy value of the cluster head is 68 percent. If it drops below 60 percent, then node n5 is elected as the cluster head.

In a clustered network, due to traffic, more number of retransmissions and low energy lead to time delay in the cluster head. Time delay occurs in the cluster head side only and not in the sensor node side. The time taken to send a message is defined as time delay. It should always be low. In a clustered network, cluster head only faces the energy issue. The node that has energy issues definitely would have a high time delay (Figure 9.3).

9.1.2 Flat network

In a flat network, nodes are scattered in an environment. The sensed information will be sent to the base station or sink node directly.

In Figure 9.4, it is illustrated that nodes n1, n2 ... n10 are scattered and all the nodes are sending their own information to the sink node. Sending the information and sensing the environment are of two categories: time based and event based.

Time based means every 30 minutes or in any prescribed time, the sensor will sense the environment and send the information to the sink node. An event-based sensor will monitor and send the information to the sink node only if the specified event occurred. For habitat monitoring, or earthquake or tsunami indications, the sensors will only send information when a prescribed threshold is exceeded.

In Figure 9.5, the energy value of each node is illustrated. In a flat network, most of the energy-affected nodes are intermediate nodes and nodes closer to

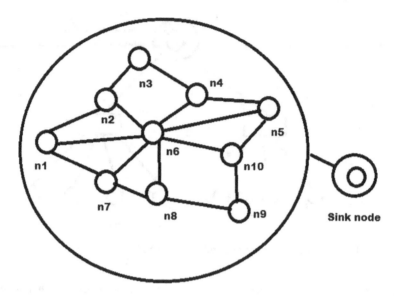

Figure 9.3 Time delay value of sensor node and cluster head.

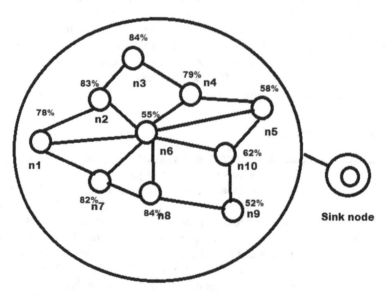

Figure 9.4 Flat network.

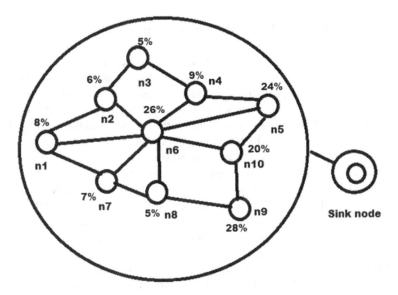

Figure 9.5 Energy value of each node in a flat network.

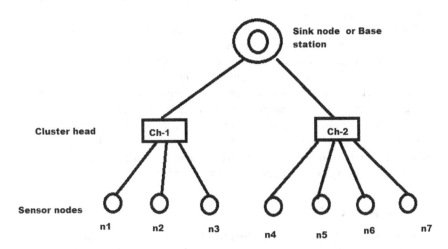

Figure 9.6 Time delay value of each node in a flat network.

the sink node. Intermediate nodes are non-leafy nodes, and other nodes are leafy nodes. Non-leafy nodes lose their energy quickly compared to the leafy nodes. In Figure 9.6, the nodes n6, n9 and n10 lose their energy quickly compared to the other nodes.

9.2 RELATED WORK

Energy factor and time delay factor functions are implemented to find a route having minimum time delay and high energy nodes for forwarding a message to the sink node [1]. Mouleeswaran et al. [2] proposed an energy consumption (ECON) model to save the energy of a node. This model will filter the same message transmitted repeatedly. This model is suited for a clustered network. It will also increase the total network's lifetime [3]. Transmission of very high data is done in a secure and efficient way by DWT for barcode modulation in handheld devices to transmit the very high data through DPSKOFDM.

Chunliang Zhou and Ming Wang [4] proposed a new model which solves the contradiction between service quality and survival time of Wireless Sensor Network (WSN). A model proposed by an algorithm SFLA reduces long large dependence of the signal. Feng et al. [5] proposed a clustering algorithm called segment equalisation clustering based on cluster head energy consumption (SECHEC). It effectively improves network lifetime and ensures the availability of the system within its entire lifespan.

Bougera et al. [6] describe an energy consumption model based on LoRa and LoRaWAN, which allows estimating the consumed power of each sensor node element. The definition of the different node units is first introduced. Then, a full energy model for communicating sensors is proposed. This model can be used to compare different LoRaWAN modes to find the best sensor node design to achieve its energy autonomy.

Damaso et al. [7] present an approach for evaluating the power consumption of WSN. Using the programming language code, they generated a consumption model which detects the power consumption of WSN applications.

Astakhova and Verzun [8] modelled the interaction process of Internet of Things devices by determining the impact on energy consumption of probability-energy characteristics. The study allowed to determine the interdependence of energy and information parameters of the sensor network. The constructed model will allow to estimate the lifetime of the sensor network. A possible development of the presented research may be the development of routing protocols that use the assessment of energy consumption by sensory devices.

9.3 ENERGY LOSS

Energy is one of the important factors for the sensor network. The longevity of the whole network is based on the energy of the sensors. Without knowing the energy of the neighbouring sensors, a sender node can send packets to them, and then the message will be lost at the neighbour itself without reaching the sink node.

A number of factors contribute to energy loss such as retransmission, monitoring and being close to the sink node. In Figure 9.7 (flat architecture), there are n numbers of nodes from n1 to n10. Based on the factors, energy-losing nodes

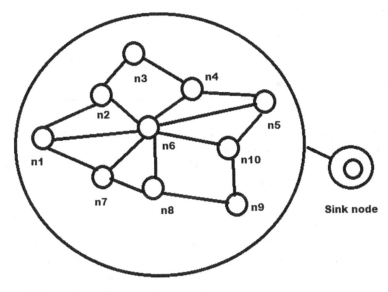

Figure 9.7 Energy loss in flat network.

are n6, n5, n9 and n10. We can divide the above nodes into leaf nodes and non-leaf nodes. Leaf nodes are located in end place of the environment. They will monitor the environment and send the information to the neighbouring node. Non-leaf nodes are present in between the nodes, that is, intermediate node. For monitoring the environment, all nodes will lose their energy equally, but the non-leaf nodes act as an intermediate node. This node receives and sends the same information many times. Hence energy loss in these nodes is high when compared to leaf nodes. From Figure 9.7, it is observed that n5, n6, n9 and n10 are going to lose their energy.

The cluster head has to decide which node is reliable for packet forwarding. Already cluster head is also facing energy issues.

9.4 PROPOSED METHODOLOGY

Routing algorithm and routing table are important in this research. Based on the routing table, all the sensors will work efficiently.

Each node in a network can be examined by sending hello packets by the neighbouring node. On receiving the hello packet, sensor will forward that to other side neighbours too. The hello packet will store the information about the sensor time delay and total amount of energy a node has.

Hello packet will come back to the sender; the sender can store the information about the nodes in the network. That sender node will share the routing table

Table 9.1 Routing table

Node-id	Energy	Time delay	Route
N1	45%	25%	D
N2	50%	10%	G
N3	85%	5%	E

information to the neighbouring node. Or else it will flood the routing table information to all other nodes in the network.

Table 9.1 is a routing table. Every node in the network has a table like this. In this table, there are four fields.

Node-id: Identifier of the node; it will be given by the base node or sink node during installation.

Energy: Available energy level of a node in a network. It should be high.

Time delay: How long time it takes for a node receiving a packet or message to forward the message or packet to the other node. Total time taken to transmit a message to other node. It should be low.

Route: Based on the characters in the router column, a particular node is used for packet routing. These characters are assigned for each node to route a packet. If energy is high and time delay is low, then that node is named as E. To calculate this route column, we used a fuzzy logic disjunction union model, which identifies the high-level energy and low time delay nodes.

Energy is the one of the important factors for a sensor node. Without energy, a node cannot do anything. Based on the energy only, a node can be efficient. This energy is wasted due to monitoring and retransmitting messages. By having a low energy, a node send a message on time. Timing is an important factor for sensor networks.

For IoTs used to detect a forest fire or habitat monitoring, timing is an important factor.

A node's low energy will reflect in time delay also. Time delay means a node takes time to send a message. That should be always low. Low time delay node will work efficiently.

Hence our fuzzy logic model will aggregate the nodes in the network by means of high energy value and low time delay value.

9.4.1 Fuzzy logics

Each node will calculate route path by using the energy value and time delay value of a node in a network. For calculating route column in the routing table, here we applied member function of fuzzy logics.

9.4.1.1 Disjunction union model

This model aggregates the high energy node and low time delay node in a wireless sensor network. Max() function will find a node having high energy. Min() function will find low time delay node from the values retrieved from max function. This disjunction union model will find the reliable node in a route and grade that node as "E". Then the node with lesser value than E will be graded as "G". The next level nodes are graded as "D".

Based on hello packets, each node can store the other node's information and update its routing table. Using max () and min() fuzzy functions, we can calculate the maximum energy level node and minimum time delay node. We can find the overall max(min()) function for any one or two nodes in a route. This node is named as the excellent node "E". Then the remaining nodes are compared with the excellent node and are named as per the max() and min() values.

We can find out the energy and time delay of the nodes based on these functions. Then the nodes are named in the three categories mentioned above plus one fault or dead node. Based on the routing needs, the nodes can be applied for the packet routing. If the information is very urgent, the excellent node can be used to save network lifetime. If a normal message is to be transmitted, we can use a combination of good and normal nodes.

Based on the information, packet routing will be changed. In this process, malicious nodes are also detected (Figure 9.8).

Transmission will be fast when compared with the other models. Based on the type of transmission, the nodes can be applied so that it reduces the amount of energy spent by all the other nodes in the network.

For urgent and most trusted information, transmission network can use the E graded nodes. For normal transmission of information, G nodes can be used. To check the node information, D nodes can be employed.

By applying the union disjunction model, the reliable nodes in flat and clustered networks are shown in Figure 9.9.

By comparing energy and time delay value of each node in a clustered architecture, the nodes are labelled as E, G and D. Using this model, the next cluster head can also be selected. In cluster-1, there is only one reliable sensor and one G sensor. In cluster-2, there are two D sensors, one G and one E sensor.

Based on the packet information's importance and urgency, the cluster head will select the node.

Figure 9.10 clearly illustrates the reliable sensor in the network. Using the disjunction union model, we can group the sensors into three categories. Based on these three categories, packets will be routed in the network. The model is also compared with other models. The union disjunction model increases the network lifetime and also the packet delivery rate. Using this disjunction union model, we can detect the reliable nodes in the IoT network.

By using the disjunction union model, the overall network time is also increased when compared with other models applied to increase the network lifetime.

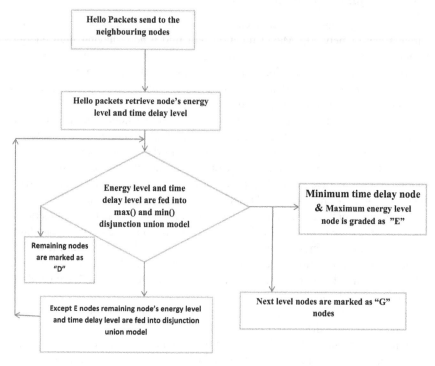

Figure 9.8 Disjunction union model.

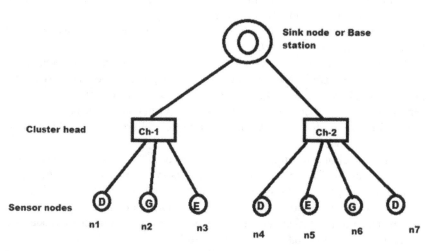

Figure 9.9 Using disjunction model reliable sensors are extracted in clustered architecture.

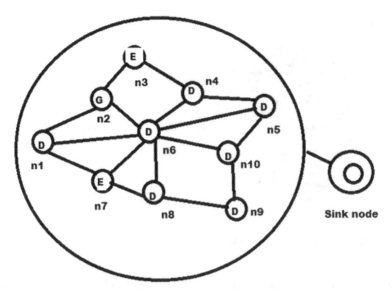

Figure 9.10 Using disjunction model reliable sensors are extracted in flat architecture.

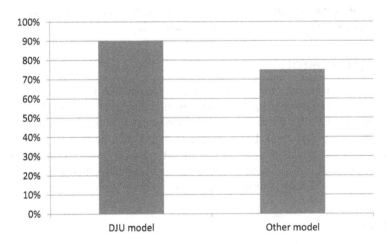

Figure 9.11 Network lifetime.

The disjunction union model works efficiently for packet routing. Based on the information forwarding, a node can select the neighbouring node and create a path. For most urgent and important information, a node can create a reliable path by selecting reliable node called "E" node in a path. E node has a more energy

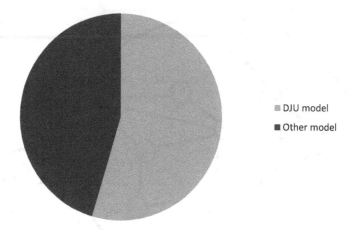

Figure 9.12 Efficiency in packet routing.

and very less time delay; hence the packet can be forwarded very fast to the destination or to the sink node. A combination of E and G nodes is also be selected to send a packet. If the situation arises that there is only one E node in a path and that it cannot reach the destination, the sending node can use E and G nodes. A combination of G and D nodes can be used for communication purposes like sending hello packets to know about the status of each node.

9.5 CONCLUSION

The disjunction union model locates the reliable node in a wireless sensor network. This model can be applied in both hierarchical and flat networks. It works efficiently in both the architectures. It improves network lifetime and maintains good packet delivery ratio. Based on the routing table, the nodes can select the path to forward a packet. For urgent purposes, E nodes are used. For normal work, G nodes are used. If there is no path to send the message to other nodes, messages are sent to the sink node. For this purpose, one or two D nodes can be used.

References

[1]   M.S. Nidhya and R. Chinnaiyan. "Conniving Energy and Time Delay Factor to Model Reliability for Wireless Sensor Networks." IEEE Xplore Digital Library, 2017.
[2]   S.K. Mouleeswaran, M.S. Nidhya and G.G. Gokilam. "Improving Energy Consumption of a Node in Wireless Sensor Networks." *International Journal of Recent Technology and Engineering* 2019, Volume 8 Issue 4. www.ijrte.org/wp-cont ent/uploads/papers/v8i4/C6478098319.pdf
[3]   Neerudu Uma Maheshwari. "Efficent Data Transmission with Barcode Modulation based on DPSKOFDM." *International Journal of Engineering and Computer Science* 2017, doi:10.18535/ijecs/v612.11

[4] Chunliang Zhou and Ming Wang. "A Wireless Sensor Network Model Considering Energy Consumption Balance." *Mathematical Problems in Engineering* 2018. doi. org/10.1155/2018/8592821

[5] Feng Luo, Chunxiao Jiang and Hajiun Zhag. "Node Energy Consumption Analysis in Wireless Sensor Networks." 2014 IEEE 80th Vehicular Technology Conference.

[6] Bouguera, Taoufik, Jean-François Diouris, Jean-Jacques Chaillout, Randa Jaouadi, and Guillaume Andrieux. "Energy Consumption Model for Sensor Nodes Based on LoRa and LoRaWAN" Sensors 18, no. 7: 2104. 2018, https://doi.org/10.3390/s18072104

[7] Antonio Damaso, Davi Freitas and Nelson Rosa. "Evaluating the Power Consumption of Wireless Sensor Network Application using Models." *Sensors* 2013, doi:10.3390/s130303473.

[8] Tatyana Astakhova and Natalya Verzun. "A Model for Estimating Energy Consumption Seen When Nodes of Ubiquitous Sensor Networks Communicate Information to Each Other." CEUR-WS.org/vol-2344/paper5.pdf.

Chapter 10

Data analytics techniques and tools in smart city applications

P. Velmurugan, A. Kannagi, M. Varsha, and K. Velusamy

CONTENTS

DOI: 10.1201/9781003335801-10

10.1 INTRODUCTION

Rapid progress in a variety of scientific and technological fields results in a massive amounts of data that need to be stored in databases. These datasets are usually large and complex, with many variables. This huge amount of data offers little useful information without processing. Because manual processing of huge information is beyond human capabilities, computational technologies are applied to develop a model that can achieve the desired purpose. Learning how to assess and extract meaningful meaning from our enterprise's digital insights is one of the major key success factors in the data-rich era. The goal of data analysis is to extract important information from data and make decisions based on that knowledge. However, data analysis is also an iterative method that includes both data gathering and data investigation because data generation is a constant activity. One of the most important aspects of data analysis is ensuring data integrity. Depending on the business and the objective of the analysis, there are a multiplicity of approaches for doing it [1].

This evidence may be leveraged to develop more effective marketing campaigns and more personalized product and service offerings. The key to increasing productivity, efficiency, and revenue is to use data analytics. Businesses can employ data analytics to better understand the trends and behaviors of their client base [2]. The results of studying data sets will indicate where they can improve, what can be computerized, which parts can be made more efficient, what is unproductive, and what should be delegated resources. Cost efficiency is upgraded because areas that are squandering a company's funds unnecessarily may be recognized, and decisions can be made on technologies that can be carried out to reduce the luxurious operations and production [3].

There are various ways data analytics may be beneficial, and the how relies largely on the organization, but at its core, it's all about helping that party in making the best decisions possible to bring their business to where it needs to be. This also includes any problems that develop, big or small, that could result in the organization losing money, among other things. Data analytics can help a company recognize problems and forecast their probable waves. As a result,

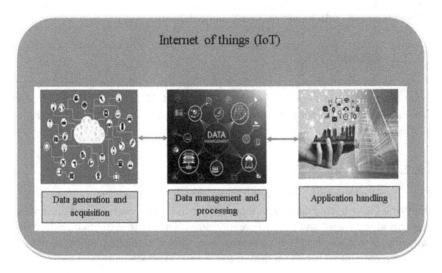

Figure 10.1 Role of IoT in smart city environments.

when used correctly and guided by professional data analysts, those data sets can be quite useful in helping a company in making educated business decisions and justifying losses [4].

10.2 STEPS OF DATA ANALYTICS

To achieve more efficient performance, the data analytics process employs five main methodologies. Identify real-time data, collected from various sources, clean inconsistent data, and analyze the data for report preparation, including interpretation [5].

10.2.1 Identify

During the identify phase, a lot of questions are asked such as what is the company's aim in terms of answering a problem, what will you use to measure your progress and what method will you use to assess it?. This is the procedure for determining data requirements or how data is organized. Gender, age, income, and demographics can all be used to separate data. Data may be grouped either numerically or category-wise.

10.2.2 Collect

Collecting the data is the second stage in data analytics. You'll need to answer the question you've identified in the raw data sets to find the relevant data set.

This can be accomplished using a variety of tools, including computers, cameras, online resources, individuals, and environmental sources.

10.2.3 Clean

A crucial factor is preparing the data for analysis. Eliminating duplicate and inconsistent data, resolving variations, regulating data structure and layout, and dealing with white spaces and other structural problems are all common examples of what this entails.

10.2.4 Analyze

The data is processed by employing using various techniques and tools for data analyzing the collected data. Developments, correlations, outliers, and dissimilarities can all be found and used to create a story. During this step, you might use data mining to discover trends in databases or data visualization software to help transform data into a graphical design that's easy to understand.

10.2.5 Interpret

The findings of your investigation is used to assess how well the evidence answered your original query? What assumptions can you make based on the data? What constraints do your conclusions represent?

10.3 TYPES OF DATA ANALYTICS

- *Descriptive analytics*: This is a summary of what has occurred throughout a specific period. This method is the evaluation of historical data to better understand the developments of the company. The use of a variety of past data to make comparisons is mentioned.
- *Diagnostic analytics*: This concentrates on why something occurred. This requires more diverse data sources as well as some speculation. Understanding why development is established or why the problem has arisen will help you take accomplishment with your business intelligence. It keeps your team from making bad judgments, especially when it comes to the distinction between correlation and causation.
- *Predictive analytics*: This transitions to what is most likely to occur shortly. It takes the investigation a step further by using statistics, computer modeling, and machine learning to calculate the probability of various outcomes.
- *Prescriptive analytics*: Prescriptive analytics foresees what will happen when it will happen, and why it will happen. It reveals which actions are most likely to yield the best results. It allows groups to solve problems, improve performance, and take advantage of opportunities.

10.4 SMART CITY

With recent advancements in the smart city model, the concept of linking everyday things over existing networks has garnered considerable attention. The growth of conventional networks that connect zillions of associated devices gave an thrust to the IoT. The IoT has been reinforced by technological improvements in WSN (wireless sensor networks), UC (ubiquitous computing), and M2M (machine-to-machine) communication. The smart city is an urban atmosphere that utilizes various technologies like information and communication technology (ICT) to improve the performance of regular city procedures and the quality of services delivered to urban citizens [6].

10.4.1 The architecture of smart city

The bottom-up smart city architecture is shown in Figure 10.2 and it is divided into four phases: (i) sensing layer, (ii) transmission layer, (iii) data management, and (iv) application layer. Because the confidence of sensitive data is of top importance for any smart city, security modules have been included in each layer. The sensing layer is placed at the bottom of the design, which is in charge of gathering data from physical devices. In the transmission layer, data transmits to the upper layers using a variety of communication protocols. The data management layer is used to collect and store dynamic information for providing services by various applications [7].

10.4.1.1 Sensing layer

A real-life smart city is made up of massive amounts of data, complicated algorithms, data storing, and intelligent decision-making capabilities. Smart city implementations, according to experts, rely on all types of data and calculations due to their significance in decision-making. Data gathering is regarded as the most significant duty because it regulates the remainder of a smart city's activities. Data collecting, on the other hand, is regarded as the most difficult activity due to the enormous variability of data. The type of information and the context have a strong influence on the data-gathering technique and technologies. A smart city is made up of different data resulting from various city functions. The sensing or data collection layer is represented by the bottom layer in the architecture of a smart city [8].

The sensing layer is made up of smart gadgets, WSNs, and other data-gathering devices. It collects data from a wide range of sensors and gadgets. The sensing layer includes several approaches for improving data capture efficiency in various situations. Use a variety of sensing technologies, such as Bluetooth, Zigbee, and radio. The camera, sensors, radio frequency identification (RFID), actuators, and global positioning systems terminals (GPS) are mostly used sensing devices. The sensing layer collects information from a variety of physical devices and networks,

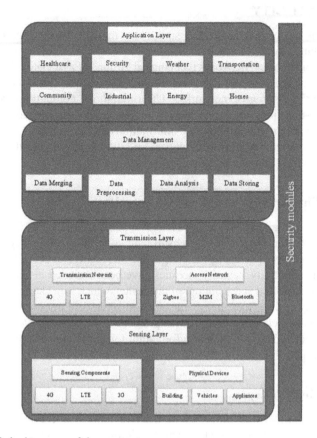

Figure 10.2 Architecture of the smart city.

increasing the number of associated devices in the network. However, more availability does not guarantee smartness or reliable data delivery. Perceptual models
are now established to correlate actions with sensors to reduce the uncertainty of
obtained data [9].

10.4.1.2 Transmission layer

The transmission layer serves as the strength of smart city architecture, linking
data sources with monitoring stations. It is a fusion of different communication
networks. Hence, routing via unique addressability is satisfied by a huge amount
of devices attached to a single network. Different types of cable, wireless, and
satellite technologies make up the transmission layer. The transmission layer
would be further separated into two sub-levels, access, and network transmission,
based on coverage. Access network technologies such as Zigbee [10], near-field

communication (NFC), Bluetooth [11], M2M [12], RFID [13], and Zwave pro-vide comparably short-range coverage. Transmission network technologies, on the other hand, are technologies that provide more coverage [14].

10.4.1.3 Data management

The data management layer resides between the sensing and application layers and is the brain of every smart city. This layer is accountable for data collection manipulation, analysis, storage, and decision-making. In reality, because the ser-vice presentation of smart city processes is dependent on data management, the efficiency of this layer is critical for a long-term smart city [15].

10.4.1.4 Application layer

The application layer is the uppermost layer, serving as a channel between citizens and the data management layer. Because it interacts directly with inhabitants, the performance of the application layer has a significant impact on user perception and satisfaction with smart city operations. Citizens are worried about the city's intelligent conduct, which provides smart services to them. Smart applications are in charge of executing decisions that are sent down from the data management layer. After receiving decisions from the data management layer, the application layer puts them into action. Because citizens are unaware of the intermediary data management layer, users' perceptions of performance improvement are purely based on the application layer's outcomes. Hence, smart systems should examine both known and unknown citizen needs and provide the best possible service while maintaining interoperability with other smart applications. Furthermore, experts recognized and asserted that incorporating modern and sophisticated technologies is insufficient to provide user happiness in actual smart cities. More study is stimulated in terms of design difficulties, optimization of necessity gathering, security views, and standards to meet these needs [16,17,18].

10.5 CHARACTERISTICS OF A SMART CITY

- Smart cities will encourage the use of technology, data, and information to expand and expand infrastructure and services. This covers the avail-ability of resources such as water and power. Providing cheap housing for all, providing enough learning and health services, and increasing IT connectivity are all priorities.
- People will be able to access a greater number of government ser-vices. Services will be available online, allowing for greater account-ability, transparency, and public participation. With the use of cyber tour workplaces, people will be able to voice their ideas and receive feed-back, as well as monitor programs and activities.

- Increased public transit accessibility, as well as innovative solutions such as smart parking, intelligent management, and integrated modal mobility. Key administrative services will be located at shorter, walkable distances in smart cities, making them more pedestrian and bike-friendly.
- To make cities safer and less disaster-prone, smart cities will reconstruct or develop unstructured and unplanned urban areas such as slums. Criminal behavior will be monitored by video monitoring, and extreme security precautions will be implemented to protect women, children, and senior citizens [19,20].
- The establishment and maintenance of parks, playgrounds, and recreational centers will help to mitigate the consequences of urban noise. Living areas will be adapted to facilitate the expanding population while also improving the quality of life.
- Infrastructure will be more sustainable and environment friendly as a result of reduced waste generation and careful use of natural resources.

10.6 APPLICATIONS OF SMART CITY

Cities and metropolitan areas are home to more than 55 percent of the world's population, a figure that may rise to 70 percent in the future as urbanization accelerates and people migrate to cities in pursuit of work. However, if cities are to be energy efficient and ecologically friendly while also providing a good quality of life, they will need better planning and infrastructure. To put it another way, cities must evolve into smart cities. The following subsections cover some of the most important smart city applications.

10.6.1 Smart community

The goal of a smart community is to improve citizen contentment and well-being in cities. To optimize the benefits of a smart city, the smart community is linked to a variety of different components. A smart community brings together a significant number of smart buildings [21], water [22], and waste management systems [23]. Smart homes and other commercial infrastructures, such as offices, schools, data centers, factories, and warehouses are examples of smart buildings. Generic smart buildings include smart appliances, sensors, and dedicated software and hardware. In terms of energy management, smart buildings and green buildings have a lot in common. Green buildings, on the other hand, place a premium on lowering energy usage and lowering carbon emissions. Most significantly, smart buildings' data-driven decision-making capability increases energy efficiency while lowering operational costs. Smart buildings, on the other hand, are about more than just energy savings. Smart buildings are also linked to other components to manage security, lighting control, surveillance, and automated operations, among other things. Furthermore, in today's world, smart waste management is seen as an essential component of any smart community. Due to rapid urbanization and industrialization, waste generation is expanding at an exponential rate. Public city services, private offices, and people all generate waste, which is properly managed through smart waste management. Waste management is divided into four stages: waste collection, disposal, recycling, and recovery. Waste management is critical for the long-term viability of smart cities, as improper waste disposal has negative consequences for human health and the environment [24].

10.6.2 Smart transportation

From the dawn of history, transportation has been a requirement for humans. This need has been extended to sea transportation, air transportation, and train transportation as a result of technological advancements. The world's traditional transportation modes were neither interconnected nor intra-connected. The concept of connecting ordinary things, on the other hand, has changed traditional transportation networks, transforming them into interconnected modern transportation systems. Consequently, numerous communication and navigation technologies are embedded in current transportation mediums. Hence, each particle of a specific transport type is linked to the others. Diverse transport mediums are interconnected to give a worldwide transportation system by extending links within the same medium [25].

With the concept of intelligent transportation systems (ITS), vehicular ad hoc networks (VANETs) have garnered a lot of attention [25]. VANETs have recently been popular for managing traffic congestion in metropolitan suburbs via vehicle-to-infrastructure (VI) and vehicle-to-vehicle (VV) communication [26]. Transport systems become capable of functioning efficiently based on data due to communication capabilities. Furthermore, smart transportation systems provide information on street congestion levels, alternative modes, and alternate routes transportation. Furthermore, smart transportation systems impose safety and security measures for travelers and walkers, as well as performance improvement [27]. Modern systems, as a result, provide global airline hubs, intercity train networks, intelligent road networks, metro train and subway, protected cycle routes, safety-enhanced public transportation, and protected pedestrian paths [28].

10.6.3 Smart healthcare

In today's society, the exponential population development rate poses several health care challenges. Hence, traditional medical methods are no longer adequate to meet the health care needs of the global population, rendering them outmoded and invalid [29]. The situation is worse since the medical practitioner in the health care sector does not keep pace with the population. Raise the chances of giving the wrong treatment, getting the wrong diagnosis, and misunderstanding the diseases of an infectious and epidemic. The hole between health care expectations and reality is widened by a lack of resources and excessive response. Smart

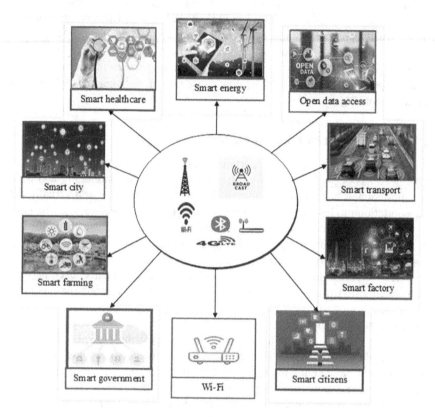

Figure 10.3 Applications of smart city.

health care methods were developed as a method to close the gap between health care demand and supply while retaining efficiency, accuracy, and long-term viability [30].

Smart energy and smart health care are defined as the fusion of conventional medical practices with modern medical intervention methods, such as medical equipment, wearable device, sensors, ICT, and emergency services. Contemporary intelligent health care services use sensor networks, smartphone applications, fog computing, cloud computing, ICT, and strong data processing techniques to meet demand and improve service quality. Smart health care facilities [31] make sensitive patient data available to authorized users such as doctors and nurses technicians via a secure hospital system network, allowing them to make real-time decisions about their patients' health. Furthermore, centrally maintained electronic health records (EHR) enable immediate decision-making based on the most up-to-date data.

10.6.4 Smart energy

Highly varied energy sources can be renewable or nonrenewable. Renewable sources such as wind, solar, and geothermal, unlike nonrenewable sources such as fossil fuel, do not decline with usage due to the nature of regeneration. Smart energy has been championed by professionals for decades [32]. To create awareness and popularize energy consumption best practices, green energy and sustainable energy concepts will be used. Green energy strives to consume energy with as little environmental impact as possible [33]. The ultimate goal of sustainable energy is to keep nonrenewable energy sources available for present and future generations to consume. Among other things, the smart energy concept appeals to people because it advocates a holistic strategy that combines sustainable energy, green energy, and renewable energy [34]. Smart energy attempts to meet energy demands while integrating renewable energy sources to ensure the long-term viability of nonrenewable energy sources and limit negative environmental effects, such as lowering carbon emissions. Renewable energy sources, as previously stated, are good fits for meeting the world's energy demands due to their renewable nature. Renewable energy's popularity has grown in tandem with rising energy demands over the last few decades. As a result, numerous researches to integrate renewable energy sources into smart buildings have been done. Renewable energy sources are connected to smart buildings in certain circumstances, and renewable energy facilities are combined with smart grids in others.

10.7 DATA ANALYSIS TECHNIQUES FOR SMART CITY APPLICATIONS

Data has become a vital commodity that is being regarded as a link between our past and future, from data analysis to intelligent machines. Cities can discover patterns and needs by analyzing data from IoT devices and sensors. While society has progressed in terms of technology, it is still unaware that data is the foundation for all of the technological breakthroughs that have combined to make the globe so evolved. When it comes to data, a variety of tools and strategies are used to arrange, organize, and collect data in the way that the user desires [35].

Smart cities use technology to improve public transportation, traffic management, water and power supply, and law enforcement, schools, and hospitals, among other things. Cities can discover patterns and needs by analyzing data from IoT devices and sensors. There are various methods have been applied to handle the data that are produced by the sensors. Hence, the following subsections are discussed the various techniques and tools for analyzing the data [36].

10.7.1 Cluster analysis

Data points are sorted into distinct clusters depending on their similarity. Objects that may be comparable are kept together in a group with limited or dissimilarity to one another. It completes this by recognizing comparable patterns in the unlabeled data set, such as size, shape, color, and movement, and classifying them according to the absence or presence of those patterns. It is an unsupervised learning method that the algorithm receives no supervision and works with an unlabeled dataset. Following the application of clustering, each cluster or group is given a cluster-id, which the algorithm can utilize to facilitate the processing of huge and complex data sets [37]. The k-means is a well-known clustering algorithm for capturing noise data in smart cities [38].

10.7.2 Cohort analysis

Cohort analysis is an examined and relates a specific section of user activity, and can be categorized with others with similar features using historical data [39]. It is practicable to get a wealth of insight into client wants or a good acceptance of a superior target group by employing this data mining process. Cohort analysis may be moderately valuable in advertising because it permits you to understand the effects of your upgrades on certain client segments. When employing the cohort analysis method, Google Analytics is a fantastic place to start.

10.7.3 Regression analysis

Regression analysis uses historical data to figure out how more than one independent variables influence the value of a dependent variable. By analyzing the relationship between each component and how it evolved in the past, you can predict different outcomes and create smarter business choices in the future [40].

10.7.4 Neural networks

The artificial neural network (ANN) is the base for machine learning's intelligent approaches. It's data-driven analytics that seeks to establish how the human brain works and forecasts values with the least amount of intervention possible. Since ANN learns from every data interaction, they evolve and improve over time. Predictive data analysis is a common use for neural networks [41]. The predictive analytics tool from data pine is one example of a BI reporting tool that has this feature built-in. Users can use this tool to quickly and easily make a variety of forecasts. Simply pick the data to be processed based on your key performance indicators (KPIs), and the software will produce estimates based on current and historical data for you. Anyone in your organization can operate it because of its user-friendly interface; you don't need to be a data scientist to do so.

10.7.5 Factor analysis

Factor analysis expresses the variation among related variables in terms of a smaller number of unknown characteristics known as factors. The aim is to discover independent latent variables, and it is an excessive way to streamline specific data parts. A factor is a collection of observed variables that respond to an activity in a comparable way. Because the number of variables in a data set can be overwhelming, Factor analysis condenses them into a smaller number of variables that are more actionable and meaningful to work with. It is a statistical strategy for reducing the number of observable variables to gain a deeper understanding of a data set [42].

10.7.6 Time-series analysis

A statistical tool for detecting trends and cycles during time is called time-series analysis. A set of data items measuring the same variable at various times is called time-series data. Analysts can calculate how the variable of interest will fluctuate in the future by looking at time-related trends. Time-series analyses are widely employed in a range of industries, with stock market analysis, economic, and sales forecasting being the most frequent. The moving average (MV), autoregressive (AR) models, and the integrated (I) models are the three main types of these models [43].

10.7.7 Sentiment analysis

Sentiment analysis is a highly helpful qualitative approach that falls within the larger category of text analysis – the (typically automated) process of sorting and is capable of understanding textual material. The sentiment analysis is to identify and categorize the emotions expressed in textual data. This lets you regulate how your consumers feel about various sections of your brand, item, or service from a commercial standpoint. Different types of sentiment analysis models are available, each focusing on a different aspect of the situation. Sentiment analysis employs a variety of natural language processing (NLP) methods and algorithms that have been trained to link specific inputs with specific outputs.

10.8 DATA ANALYTICS TOOLS

Users can utilize data analytics tools to uncover patterns, trends, and connections in their data. This information assists them in making better decisions, predicting risks, and seeing possibilities. Some of the most well-known data analytics tools that are utilized in smart applications are mentioned in the subsections below.

10.8.1 R programming language

R is an extensively used statistical and data analysis tool that is open source and its features are shown in Figure 10.4. R comes with a command-line interface by default. In addition, the R programming language is the most up-to-date tool. The Development Core Team is now working on it, which was conceived by

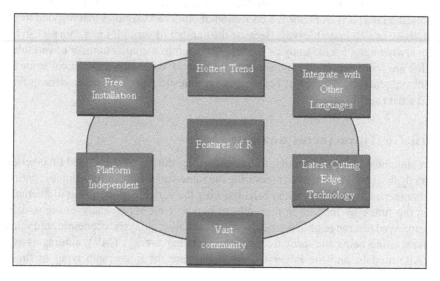

Figure 10.4 Features of R programming languages.

Ross Ihaka and Robert Gentleman at the University of Auckland in New Zealand. The R programming language is a fork of the S programming language. It also works with Scheme-inspired lexical scoping semantics. Furthermore, the project was inaugurated in 1992, with a first version launched in 1995 and a beta version delivered in 2000. Various applications have been solved by using R in smart city applications [44].

10.8.2 Python

Python is one of the most popular open-source, high-level programming, interpreted language with a simple syntax and dynamic semantics. Python is a completely free programming language that can be taken from the official website. Python is used by multinational corporations such as YouTube, Facebook, and Facebook. Python is improving its features and functionalities to make data analysis more efficient and accurate. That is, new releases with updated features are released regularly. All machine learning algorithms and artificial intelligence-based algorithms have been solved in various applications by using the Python programming language.

10.8.3 Spark

Apache Spark is a set of libraries and a unified computing engine for parallel information processing on computer clusters. Spark is an open source for this

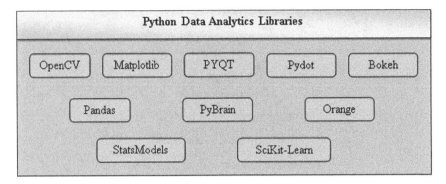

Figure 10.5 Python data analytics libraries.

task at the time of this writing, making it the de-facto tool for any data scientist and developer interested in big data. Spark runs on anything from a CPU to a cluster of thousands of servers and supports numerous widely used programming languages such as Java, Python, R, and Scala. It also includes libraries for a variety of tasks ranging from SQL to streaming and machine learning. This makes it a simple system to get started with and scale up to massive big data processing. Figure 10.5 shows the libraries of Python for data analytics [45].

Spark is developed to handle a wide range of data analytics activities, ranging from simple data loading and SQL queries to machine learning and streaming calculation, over the same computational engine and with a consistent set of APIs. Spark has libraries for SQL and structured data called Spark SQL, machine learning called MLlib, stream processing called Spark Streaming, and graph analytics called Spark Graph Analytics. Hundreds of open sources are available such as external libraries, ranging from connectors for various storage systems to machine learning techniques, in addition to these libraries.

10.8.4 Tableau

It's a popular business intelligence application tool for analyzing and visualizing data in an easy-to-understand style [46]. Tableau, which has been awarded a leader in the Gartner Magic Quadrant for data analysis for the eighth year, allows users to work on live data sets and spend more time on data analysis rather than data wrangling. It's a full-featured platform that lets you prepare analyses, collaborate, and share big data insights. Tableau is a leader in self-service visual analysis, permitting users to ask new questions of managed big data and quickly share their findings across the company. Tableau is free that anyone may use to create visualizations, but you must save your worksheets or workbook to the Tableau Server, which anyone can view. Figure 10.6 shows the advantages of the tableau.

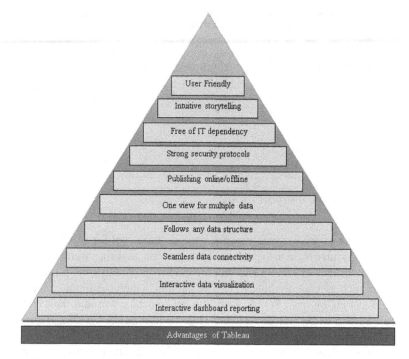

Figure 10.6 Advantages of Tableau.

10.8.5 Konstanz Information Miner (KNIME)

The KNIME is open-source data analytics software that allows for data integration, processing, visualization, and reporting [47]. It incorporates data mining and machine learning libraries with little to no code. KNIME is a great tool for data scientists who don't have a lot of programming experience but need to incorporate and analyze data to develop machine learning and other statistical models. Its graphical interface allows for easy analysis and modeling with just a few clicks.

- KNIME Analytics Platform: Open-source platform is used to clean and collect the data, make reusable modules accessible to everybody, and construct data science workflows.
- KNIME Server: This is a platform for deploying data science workflows, team communication, management, and automation in companies.

10.8.6 Rapid miner

As a result, each of the tools' durable components is simple to use [48]. The visual workflow designer, which is a tool that offers users a visual environment,

Figure 10.7 Process of Rapid Miner.

is one of the system's features. Analytics processes can be built, created, and deployed in this environment. Models and visual presentations can also be created and handled here. Because of the comfortable environment, users may effortlessly complete all of these tasks. Figure 10.7 shows the process of the rapid miner.

10.8.7 Tensorflow

With Google Brain, you can use Tensorflow, an open-source toolkit for numerical and large-scale computing, to make gathering data, training models, serving predictions, and refining future outcomes easier [17]. It combines deep learning and machine learning algorithms into a single package. It makes use of Python as a comprehensible front-end and runs it in optimized C++. It is frequently utilized in modern technologies such as AI, data science, and machine learning.

10.9 CHALLENGES OF DATA ANALYTICS IN SMART CITY APPLICATIONS

Smart technologies and real-time data can be used to better utilize infrastructure, services, and energy. Sensors and other interconnected systems collect information and data about the smart city, which is communicated over wired or wireless networks. These data are processed and analyzed to increase a better understanding of current events and create predictions about what will happen in the future. Smart cities generate an enormous amount of data, which will only increase in the future. The use of data analytics in smart cities is highly problematic. Smart cities produce a massive amount of data, which causes a challenge for big data analytics. The volume of data exceeds the capacity of typical databases or data warehouses, necessitating the adoption of alternative technology to manage petabytes or exabytes of data.

The velocity of data is quite high since the incoming data is flowing in at a faster rate, which causes data aging; it is difficult to determine the value of data.

Similarly, sensor data accuracy might lead to deceptive analytics due to the data's lack of credibility. There's also the problem of data privacy, which could contain personal information about government officials, citizens, and service providers. New technology, such as Hadoop, is being used to address the problem of massive amounts of data. The k-anonymity is a well-famous data anonymization technique that can be used to keep sensitive and personal information about government, citizens, and service providers.

10.10 CONCLUSIONS

The proliferation of artificial intelligence, sensor technologies, and wireless connectivity has enabled ubiquitous sensing via distributed sensors. These sensors are networks in numerous sectors that lead to smart systems. The smart system's sensor technologies will generate a large amount of data. Dealing with data is a new challenge in the business world that has caught many businesses off guard. Now that we have a better understanding of the potential of collected data, we're looking for new and better ways to collect, organize, and store it. Hence, To make smart decisions, data management is an important step. The current chapter covers the basics of data analytics, smart cities, smart city architecture, and its applications, the role of data analytics in smart city applications, various data analytics tools and approaches, and challenges.

References

[1]   Camero A and Alba E 2019 Smart City and information technology: A review. *Cities* **93**, 84–94.

[2]   Fawzy D, Moussa S and Badr N 2016 The evolution of data mining techniques to big data analytics: An extensive study with application to renewable energy data analytics. *Asian Journal of Applied Sciences* **4**(3). Retrieved from https://ajouronline. com/index.php/AJAS/article/view/3792

[3]   Lai C S, Jia Y, Dong Z, Wang D, Tao Y, Lai Q H, Wong R T, Zobaa A F, Wu R and Lai L L 2020 A review of technical standards for smart cities. *Clean Technologies* **2**, 290–310.

[4]   Puiu D, Barnaghi P, Tönjes R, Kümper D, Ali M I, Mileo A, Parreira J X, Fischer M, Kolozali S and Farajidavar N 2016 Citypulse: Large scale data analytics framework for smart cities *IEEE Access* **4**, 1086–1188. doi: 10.1109/ ACCESS.2016.2541999

[5]   Silva B N, Khan M and Han K 2018 Towards sustainable smart cities: A review of trends, architectures, components, and open challenges in smart cities. *Sustainable Cities and Society* **38**, 697–713.

[6]   Song H, Srinivasan R, Sookoor T and Jeschke S 2017 *Smart Cities: Foundations, Principles, and Applications*. John Wiley & Sons.

[7]   Gavrilović N and Mishra A 2021 Software architecture of the internet of things (IoT) for smart city, healthcare and agriculture: analysis and improvement directions. *Journal of Ambient Intelligence and Humanized Computing* **12**, 1315–36.

[8] Cecaj A, Lippi M, Mamei M and Zambonelli F 2021 Sensing and forecasting crowd distribution in smart cities: Potentials and approaches. *IoT* **2**, 33–49.

[9] Haque A K M B, Bhushan B and Dhiman G Conceptualizing smart city applications: Requirements, architecture, security issues, and emerging trends *Expert Systems*. https://doi.org/10.1111/exsy.12753

[10] Gupta M and Singh S 2021 A survey on the ZigBee protocol, it's security in Internet of Things (IoT) and comparison of ZigBee with Bluetooth and Wi-Fi. In: *Applications of Artificial Intelligence in Engineering*, Springer, pp. 473–82.

[11] Hasan R, Hasan R and Islam T 2021 InSight: A Bluetooth beacon-based ad-hoc emergency alert system for smart cities. In: *2021 IEEE 18th Annual Consumer Communications & Networking Conference (CCNC)*: IEEE, pp. 1–6.

[12] Kaushik S, Aggarwal G and Tejasvee S 2021 A glimpse on key applications of smart city under M2M communication. In: *IOP Conference Series: Materials Science and Engineering*: IOP Publishing, p. 012006.

[13] Santoso B and Sari M W 2021 Developing parking queue monitoring system using Wireless Sensor Network and RFID technology. In: *Journal of Physics: Conference Series*: IOP Publishing, p. 012056.

[14] Jiang Y 2021 Economic development of smart city industry based on 5G network and wireless sensors. *Microprocessors and Microsystems* **80**, 103563.

[15] Chen J, Ramanathan L and Alazab M 2021 Holistic big data integrated artificial intelligent modeling to improve privacy and security in data management of smart cities. *Microprocessors and Microsystems* **81**, 103722.

[16] Ghotbou A and Khansari M 2021 Comparing application layer protocols for video transmission in IoT low power lossy networks: an analytic comparison. *Wireless Networks* **27**, 269–83.

[17] Ahmed, S T, Sreedhar Kumar S, Anusha B, Bhumika P, Gunashree M. and Ishwarya B (2018, November). A generalized study on data mining and clustering algorithms. In *International Conference On Computational Vision and Bio Inspired Computing* (pp. 1121–1129).Cham: Springer.

[18] Ahmed S T, Singh D K, Basha S M, Nasr E A, Kamrani A K and Aboudaif M K (2021). Neural network based mental depression identification and sentiments classification technique from speech signals: A COVID-19 Focused Pandemic Study. *Frontiers in Public Health*, 9.

[19] Periasamy K, Periasamy S, Velayutham S, Zhang Z, Ahmed S T and Jayapalan A (2021). *A Proactive Model to Predict Osteoporosis: An Artificial Immune System Approach*. Expert Systems.

[20] Sathiyamoorthi V, Ilavarasi A K, Murugeswari K, Ahmed S T, Devi B A and Kalipindi M (2021). A deep convolutional neural network based computer aided diagnosis system for the prediction of Alzheimer's disease in MRI images. *Measurement* **171**, 108838.

[21] Mlýnek P, Rusz M, Benešl L, Sláčik J and Musil P 2021 Possibilities of broadband power line communications for smart home and smart building applications *Sensors* **21**, 240.

[22] Singh M and Ahmed S 2021 IoT based smart water management systems: A systematic review. *Materials Today: Proceedings* **46**, 5211–8.

[23] Roshan R and Rishi O 2021 Effective and efficient smart waste management system for the smart cities using Internet of Things (IoT): An Indian perspective. *Rising Threats in Expert Applications and Solutions* 473–9.

[24] Chen D, Wawrzynski P and Lv Z 2021 Cyber security in smart cities: a review of deep learning-based applications and case studies. *Sustainable Cities and Society* **66**, 102655.

[25] Sundaresan S, Kumar K S, Nishanth R, Robinson Y H and Kumar A J 2021 *Blockchain for Smart Cities*. Elsevier, pp. 35–56.

[26] Li B, Kisacikoglu M C, Liu C, Singh N and Erol-Kantarci M 2017 Big data analytics for electric vehicle integration in green smart cities *IEEE Communications Magazine* **55**, 19–25.

[27] Castellanos C, Perez B and Correal D 2021 *Smart Cities: A Data Analytics Perspective*: Springer, pp. 161–79.

[28] Rodríguez-Bolívar M P 2015 *Transforming City Governments for Successful Smart Cities*. Springer.

[29] Hossain M S, Muhammad G and Alamri A 2019 Smart healthcare monitoring: A voice pathology detection paradigm for smart cities. *Multimedia Systems* **25**, 565–75.

[30] Ghazal T M, Hasan M K, Alshurideh M T, Alzoubi H M, Ahmad M, Akbar S S, Al Kurdi B and Akour I A 2021 IoT for Smart Cities: Machine learning approaches in smart healthcare – a review. *Future Internet* **13**, 218.

[31] Sheng T J, Islam M S, Misran N, Baharuddin M H, Arshad H, Islam M R, Chowdhury M E, Rmili H and Islam M T 2020 An internet of things based smart waste management system using LoRa and Tensorflow deep learning model. *IEEE Access* **8**, 148793–811.

[32] Kim H, Choi H, Kang H, An J, Yeom S and Hong T 2021 A systematic review of the smart energy conservation system: From smart homes to sustainable smart cities. *Renewable and Sustainable Energy Reviews* **140**, 110755.

[33] Zhang X, Manogaran G and Muthu B 2021 IoT enabled integrated system for green energy into smart cities. *Sustainable Energy Technologies and Assessments* **46**, 101208.

[34] Konstantinou C 2021 Towards a secure and resilient all-renewable energy grid for smart cities *IEEE Consumer Electronics Magazine* 11, no. 1, pp. 33–41, 1 Jan. 2022, doi: 10.1109/MCE.2021.3055492

[35] Khan M A, Algarni F and Quasim M T 2021 *Smart Cities: A Data Analytics Perspective*. Springer. https://link.springer.com/book/10.1007/978-3-030-60922-1

[36] Quasim M T, Khan M A, Algarni F and Alshahrani M M 2021 *Smart Cities: A Data Analytics Perspective*. Springer, pp. 3–16. https://link.springer.com/book/10.1007/978-3-030-60922-1

[37] Tharwat M and Khattab A 2021 *Smart Cities: A Data Analytics Perspective*: Springer, pp. 113–34. https://link.springer.com/book/10.1007/978-3-030-60922-1

[38] Ran X, Zhou X, Lei M, Tepsan W and Deng W 2021 A novel k-means clustering algorithm with a noise algorithm for capturing urban hotspots *Applied Sciences* **11**, 11202.

[39] Kanaga E G M and Jacob L R 2021 *Intelligence in Big Data Technologies—Beyond the Hype*. Springer, pp. 97–106.

[40] Soomro K, Bhutta M N M, Khan Z and Tahir M A 2019 Smart city big data analytics: An advanced review. *Wiley Interdisciplinary Reviews: Data Mining and Knowledge Discovery* **9**, e1319.

[41] Allam Z 2019 Achieving neuroplasticity in artificial neural networks through smart cities. *Smart Cities* **2**, 118–34.

[42] Habib A, Alsmadi D and Prybutok V R 2020 Factors that determine residents' acceptance of smart city technologies. *Behaviour & Information Technology* **39**, 610–23.

[43] Kumari A and Tanwar S 2020 Secure data analytics for smart grid systems in a sustainable smart city: Challenges, solutions, and future directions. *Sustainable Computing: Informatics and Systems* **28**, 100427.

[44] Garg R, Malik A and Raj G 2018 A comprehensive analysis for crime prediction in smart City using R programming. In: *2018 8th International Conference on Cloud Computing, Data Science & Engineering (Confluence)*. IEEE, pp 14–15.

[45] Suma S, Mehmood R and Albeshri A 2020 *Smart Infrastructure and Applications*. Springer, pp 55–78.

[46] Dameri R 2015 *From Information to Smart Society*. Springer, pp. 173–80.

[47] Jara A J, Genoud D and Bocchi Y 2015 Big data for smart cities with KNIME a real experience in the SmartSantander testbed. *Software: Practice and Experience* **45**, 1145–60.

[48] Hofmann M and Klinkenberg R 2016 *RapidMiner: Data Mining Use Cases and Business Analytics Applications*. CRC Press.

Chapter 11

Machine learning algorithms for IoT applications

R. Kavitha, K. Saravanan, S. Adlin Jebakumari, and K. Velusamy

CONTENTS

DOI: 10.1201/9781003335801-11

11.1 INTRODUCTION

On the Web, lots of network devices are connected for sharing the data called the Internet of Things (IoT). The connections of network devices are called smart devices such as sensors, actuators, mobile phones, and other smart devices [1]. It also enables these smart devices to be controlled over the internet via wired or wireless communication networks, allowing them to communicate and exchange data. The main aim of the IoT system is to offer various objects to the linked in any location anytime by anyone using any network, path, and any services. According to a statistic report published by IDC (International Data Corporation), there are over 55.7 billion IoT devices connected to the universe, which will produce over 80 ZB of data by 2025. By collecting data from IoT devices and analyzing it to gain intelligence and appreciate the environment, structures can be built to improve quality of life, namely diagnosis of machine assessment, health monitoring, human body activities, structural monitoring, and localization. As the IoT becomes more widespread and extensively used, huge amounts of data are generated by numerous devices and sensors. Many IoT applications have been developed to provide users with more precise and fine-grained services [2]. These big data from the IoT devices can be further handled and examined to provide intelligence to IoT service providers and users. Many data-driven analytic procedures are used in emerging IoT applications to efficiently use big IoT sensing data [3]. Figure 11.1 shows the progress of the IoT network from 2016 to 2026.

On the other hand, machine learning (ML) as a technique gradually evolved with its algorithms to become more robust, effective, and accurate [4]. Many real-world application domains, such as image analysis, fraud detection, disease prediction, and agriculture monitoring, have benefited from the use of ML

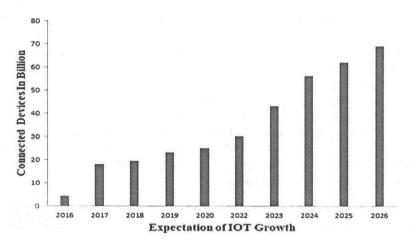

Figure 11.1 IoT growth projections from 2016 to 2026.

techniques. Recently, ML algorithms have been used to improve the performance of several IoT applications by designing algorithms that use the collected data to increase efficiency [5]. Hence, not only does machine learning enhance autonomous control in IoT applications, but it also enhances their intelligent decision-making skills. However, when it comes to IoT applications, ML techniques face numerous challenges that must be properly considered. In this chapter, we make a brief overview of the IoT, including history, architecture, and layers. We discuss various ML algorithms with their various types, such as supervised learning, unsupervised learning, and reinforcement learning.

The rest of the chapter is arranged as follows. The history of IoT networks is discussed in Section 11.2. Sections 11.3, 11.4, and 11.5 discuss the architecture, applications, and challenges of IoT networks, respectively. Section 11.6 provides an introduction to machine learning. In Section 11.7, the role of machine learning algorithms in IoT applications is discussed. The various machine learning algorithms for IoT applications are discussed in Section 11.8, and the chapter's conclusions are drawn in Section 11.9.

11.2 HISTORY OF IOT

In the 1980s, the first linked device was a Coca-Cola vending machine located at the Carnegie Mellon University and developed by four students. The main idea of the machine is to indicate whether drinks contained in the coke machine are cold or not. This development led to further studies in the field, which resulted in the improvement of connected machines all over the world. This establishment had inspired a lot of technologists all over the world to make their own connected devices that cover wider distances and use radio frequency identification (RFID) and ultra-high frequency (UHF). In the 1990s, the IBM scientist filed a patent for a faster data transfer technology. However, they were not commercialized due to financial problems. Then they sold their patent to the bar code provider called Intermec. Several enterprises recommended solutions like Microsoft's at Work or Novell's NEST between 1993 and 1997.

Norton estimated that the IoT-connected devices will be more than 21 billion by 2025. North America is expected to have 29 percent of the world's self-driving fleet by 2035. IDC (International Data Corporation) reported that there will be 55.7 billion connected devices worldwide by 2025, 75 percent of which will be connected to an IoT platform. Table 11.1 shows the development of phases of the IoT system starting from 1982 till 2021.

Table 11.1 History of IoT networks

Year	Contribution
1982	The four students invented the ARPANET (Advanced Research Projects Agency Network)–linked coke machine from Carnegie Mellon university
1989	WWW (World Wide Web) proposed by Tim Berners-Lee
1990	John Romke introduced an internet-connected toaster
1999	Kevin Ashton developed his first time IoT
2000	LG rolled out the world first internet-enabled refrigerator
2004	The IoT technique is more popular. There were many research articles and magazines published
2005	Nabaztag is an internet-based device that was appeared
2008	The first IoT international conference was conducted in Zurich, Switzerland. Cisco reported that IoT devices connected more than people population
2009	The self-driving car was started by Google using the sensor-enabled device on the car deck
2010	Chinese Premier Wen Jiabao considered the IoT as a key and important engineering field
2011	Some companies have interested the internet providers to be ready for the changeover from IPv4 to IPv6
2013	Smart glasses established by Google that has featured a display ability to show information hands-free, voice recognition in natural languages to connect to the internet with the help of spoken comments
2015	IoT-enabled Barbie toys with an enabled Wi-Fi produces by Mattel and established communicating features containing a toy over with fire and voice controlling
2016	The smart home device was developed by Apple and Google home is a third-party service to facilitate the user with a communication field
2017	Azure IoT Edge was developed by Microsoft that enabled small devices to employ cloud facilities even if they are not associated with the cloud
2018	IoT technique was developed for the security-related activities
2020	As per the Cisco report, 50 billion devices will be connected
2021	BMW, Ford, and Volvo planned to launch autonomous cars

11.3 THE ARCHITECTURE OF IOT NETWORKS

Three-layered IoT architecture contains three layers such as perception layer, networking layer, and application layer. The perception layer is the physical layer that is made up of sensors that collect data about the environment and actuators that identify actions. The network layer (transmission layer) is incapable of designing routing decisions as well as the processing and transmission of data that is obtained or transferred to the perception layer. The application layer provides an interface between the IoT network and humans by delivering application-specific services to end-users.

11.4 APPLICATIONS OF IOT NETWORKS

In a real-time context, the IoT network may connect all physical and virtual items, leading to new applications and services. There are many applications are utilized IoT networks including smart home, health care, agriculture, smart city, smart grid, smart car, industrial IoT, and wearable devices. Adopting these applications makes our daily lives easier and Figure 11.2 shows the applications of IoT networks. The parts that follow go through the numerous IoT applications.

11.4.1 Smart home

The smart home application in the IoT network is the most popular application for making our home appliances easier to use. The actuator and sensor are connected

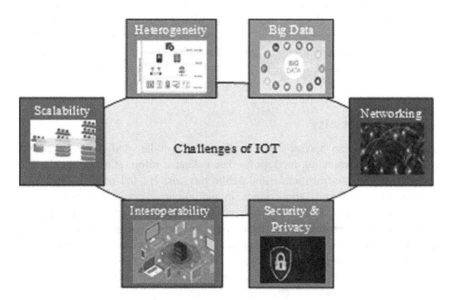

Figure 11.2 Challenges of IoT systems.

to a wireless sensor network, allowing users to connect and control a variety of household appliances. Sensors are connected to household equipment, allowing for automated or smart services to reduce human labor. For example, humans can generate massive privacy and security challenges in their homes by using cameras to connect them. The smart technology may connect the house owner at any time and provide a notification if any abnormal movements are detected [6].

11.4.2 Health care

The IoT focuses more on providing service human healthcare to prevent patient risks. Many wearable devices have been implemented and used to keep track of the patient's health. Smart wearable devices can support the monitoring of patient behavior and can assist elderly patients in living without worries. The devices can store patient progress data and send an alert notification to the doctor if unexpected changes in the patient's body are discovered. If the patient was found to be at low risk, the device advised treatment. Otherwise, the device can make an immediate call to an ambulance or send a message to the physicians [7].

11.4.3 Agriculture

Soil characteristics include humidity, temperature, and salt level, which may be collected and monitored by a sensor device to improve agriculture productivity. The remote sensing and geographic information system is a wireless technology system that may help to collect useful information about the soil in a timely and effective manner. The collected data helps in the replacement of human effort with automated technology to increase agricultural productivity. The sensor can be used to collect crop progress, soil quality, equipment efficiency, and weather conditions besides staff performance. IoT systems may assist with a variety of automated operations across the agricultural life cycle, allowing for better control of methods of production and higher crop standards of quality [8].

11.4.4 Smart city

The IoT system can operate and manage sensors, lights, and meters to collect data from the surrounding metropolis. The collected information can be used to enhance the city's infrastructure and public services. The IoT network is utilized in a variety of smart city applications, including garbage collection, smart street lighting, traffic control, and smart parking. For example, the smart traffic application can assist the general public and drivers in selecting the optimal route to save time and effort. In the event of an accident, the smart application can also notify the driver about congestion. The sensors collect the information about trash bins and can send messages to an authorization phone about the status of a trash bin for optimizing trash trucks and reduce emerge usage [9].

11.4.5 Smart grid

Using IoT sensors to gather useful information about the usage of energy in the home can help you to spend energy more effectively and save money in the long run. The smart grid approach knowledge comes from the power flow cycle from the supplier to the customer. This type of information can assist consumers in understanding dynamic pricing and electricity use. For example, a smart meter, which is used to evaluate and store information about power usage, is a well-known smart grid application. This information can assist people in lowering their energy usage and costs by altering their lifestyles [10].

11.4.6 Smart car

The smart car has internet access and can share data with other devices. The connected smart car has a lot of benefits over a regular car. By allowing the driver to operate remotely, the smart automobile can reduce accidents and driver errors. BMW, Volvo, and Ford are well-known automobile manufacturers, and they have stated that fully automated cars would be available by the end of 2021 [11].

11.4.7 Wearable devices

Wearable devices are gaining in popularity throughout the world. Several manufacturers are planning to produce more wearable devices to satisfy real-time requirements [12]. The wearable device may be able to gather data using a sensor and connect to the internet for sharing data. The data collected is saved on the Web and will be used to create useful information. Health, entertainment, and fitness are the most common data acquired from wearable devices [13].

11.4.8 Industrial IoT (IIoT)

The industrial IoT is made up of interconnected sensors, equipment, and other devices that are connected to industrial computer applications such as production, power management, and so on. Companies can detect systemic problems and pay attention to the critical issues, saving time and money while supporting business intelligence efforts [14].

11.5 CHALLENGES OF IOT

IoT environment can solve various solutions that provide countless benefits and give rise to many challenges. The general issues of the IoT networks are discussed including networking, big data, scalability, interoperability, heterogeneity, and security, and privacy. Figure 11.3 shows the challenges of IoT network systems.

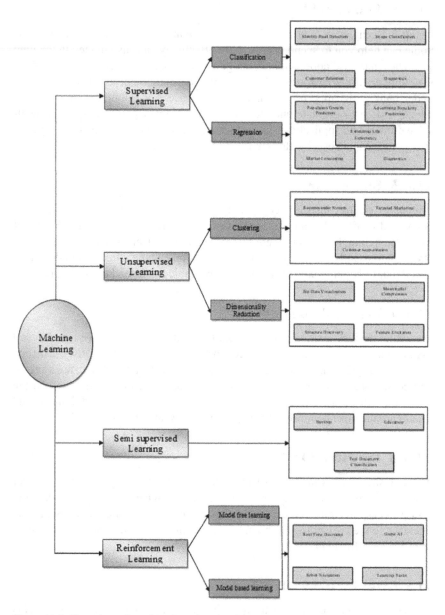

Figure 11.3 Overview of machine learning algorithms.

11.5.1 Networking

The IoT network's main goal is to connect objects or devices to share data. The connected devices have varying structures and shapes, necessitating the use of various communication networking protocols. Networking protocols for the IoT should be designed by considering system speed and performance. Network performance is determined by the networking protocol, and this issue must be addressed. Hence, choosing the right networking protocol is a common problem that can help to enhance the performance of the network.

11.5.2 Big data

The collected data may be structured or unstructured, making it difficult to process using traditional database and software methods. IoT devices collect massive amounts of data, which is referred to as big data, and handling big data can be difficult. The IoT environment is connected to billions of devices, making it a major source of big data. Even though cloud computing could be used to store information indefinitely; processing such large amounts of data is a significant challenge, especially because the performance of various IoT applications is dependent on data management services. Furthermore, big data will raise security and privacy issues because maintaining data integrity will be a challenging task.

11.5.3 Scalability

The most important issue in the IoT network is scalability, which must be addressed to keep up with the dramatic increase in devices connected. Scalability refers to a system's ability to deal with its future growth in a cost-effective manner without compromising its performance. Scalability refers to a system's ability to deal with the system's potential growth in a cost-effective fashion without compromising performance. Because people's interests change with the sense of the world, scalability is a must for the IoT system to reach differing specifications.

11.5.4 Interoperability

The potential of system components to work together efficiently, despite their technical specifications, is referred to as interoperability. IoT network devices are linked together to share information, resulting in the creation of new services. As the IoT system got popular and the number of connected devices became more, interoperability became a primary concern for proficiently connecting various objects. Hence, interoperability is another major issue that must be addressed.

11.5.5 Heterogeneity

Another challenge is the heterogeneity of the IoT network, which arises from the various devices connected, each with its own set of characteristics. The main goal of the IoT network is to establish a common technique for abstracting from the heterogeneity of such devices and trying to maximize their features. When attempting to implement a service for the IoT environment, the service provider takes into account the wide range of communication-compatible devices, protocols, and methods of communication.

11.5.6 Security and privacy

Due to heterogeneous device connectivity, another major challenge is security and privacy. Each connected devices or objects has its own set of characteristics and protocols, which have an impact on network performance. Hence, the IoT environment increases the chances of security vulnerabilities in connected devices that are badly secured and lack constructed security protocols [15]. Therefore, handling security issues is another challenge in the IoT environment, and it should be a primary concern for the IoT system to improve its adaptation to an IoT application. Also, as connected devices and related applications become more incorporated into people's daily events, IoT users must have wide-ranging confidence in their security [16].

11.6 MACHINE LEARNING

Machine learning (ML) is a kind of Artificial Intelligence (AI) that enables machines to learn without having to be programmed explicitly. ML is the process of learning to execute tasks to achieve certain goals, such as classification, clustering, prediction, pattern recognition, and so on. Computer systems are trained to archive the learning process by analyzing sample data using several algorithms and statistical models. The sample data is often analyzed using features, which are computable properties. The ML algorithm seeks to draw a connection between the features and certain output values termed as labels [17]. The data gathered during the training phase is then applied to find patterns or make decisions based on relevant content. The computational time and convergence rate are considered when choosing machine learning algorithms to solve real-world problems. The main goals of ML algorithms are to analyze data structure and integrate it into models that are easy to understand and apply. ML algorithms may be divided into four types based on the learning style such as supervised, unsupervised, semisupervised, and reinforced learning methods. The kinds of these categories are described in the next sections. Table 11.2 shows the comparison of ML algorithms and Figure 11.4 depicts a broad overview of ML techniques [18].

11.6.1 Supervised learning

Supervised learning is a machine learning technique in which computers learn from well-labeled training data and then predict output based on that data.

Table 11.2 Machine learning performance

Learning type	Input	Output	Goals	Merits	Demerits
Supervised learning	Labeled data	Known	Making predictions	High accuracy	High computation time
Unsupervised learning	Unlabeled data	Unknown	Perform the data distribution without characteristics between the actual and target variables.	Less complexity	Less accuracy
Semisupervised learning	More unlabeled data Limited labeled data	Limited known	Both supervised and unsupervised can be accomplished	It is not necessary to have a huge labeled data collection and a high level of accuracy	Labeled data is difficult to get and high computation
Reinforcement learning	Rewards	Actions	Learning focus on experience for decision-making by using reward where feedback is actions	High accuracy without human intervention	High computation time

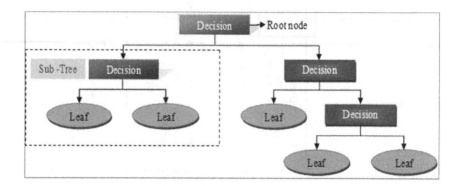

Figure 11.4 Decision tree algorithm.

Regression and classification is a type of supervised learning concept that is used to solve various real-world problems including approximating life experience, forecasting weather, and predicting population increase, digit and speech recognition, fraud detection, and diagnostics by using algorithms. Support vector machine (SVM), random forest (RF), and nearest-neighbor are examples of supervised learning algorithms. In the supervised learning technique, there are two types of phases: training and testing. The data samples considered in the training phase must have well-defined labels. The algorithms aim to predict the expected output of the testing data by learning the relationship between input values and labels [19].

11.6.2 Unsupervised learning

Unsupervised learning is a kind of machine learning in which algorithms are trained on unlabeled data and then allowed to operate on it without supervision. It deals with dimensionality reduction, the identification of hidden structures, and feature elicitation. Unsupervised learning algorithms attempt to classify patterns in test data, cluster the data, or forecast future values. The clustering and association is an example of an unsupervised learning concept. K-means clustering method is a distinguished unsupervised learning algorithm [20].

11.6.3 Semisupervised learning

The semisupervised learning approach combines the benefits of both supervised and unsupervised learning. Semisupervised learning problems are solved using both labeled and unlabeled data. It functions largely in the same way as unsupervised learning, but with the added benefit of a little amount of labeled data [21].

11.6.4 Reinforcement learning

Another type of machine learning method is reinforcement learning, which tries to predict the output based on a bunch of tuning parameters. The determined output is now being used as an input parameter to calculate new outputs until the optimal one is determined. Reinforcement learning is mostly utilized in real-time applications including AI gaming, robot navigation, skill acquisition, and real-time decision-making [22].

11.7 ROLE OF MACHINE LEARNING IN IOT APPLICATIONS

ML is a field that is constantly evolving across a range of circumstances, technologies, and disciplines, including IoT. Sensor networks, connected objects, and their applications have emerged as a result of the emergence of new technologies and developments in the field of information processing and networks. ML methods were first used to create predictive models for IoT. However, ML has shown to be a rich domain, and those who want to apply it to IoT should appreciate in ways that generate the most benefits. The use of ML techniques in an IoT network to address a variety of problems offers significant benefits in terms of precision and flexibility. IoT deployments have risen in recent years, allowing the technology to achieve widespread acceptance and appeal while also offering new challenges. These problems must be given specific attention to improving service quality [23].

More connected IoT objects have various challenges like power-saving, scalability security and privacy, long-range network, network congestion and overload, network management, interoperability and heterogeneity, QoS, and network mobility and coverage, and these should be given consideration. Most of these issues can be solved with ML, which has the advantage of allowing the network to grow without producing problems. For example, data analytics is the primary problem that must be addressed with any advancing technology. ML provides a suitable solution to predictive analysis for the information collected by the sensors such as weather prediction, disease prediction, and traffic detections.

11.8 MACHINE LEARNING ALGORITHMS FOR IOT APPLICATIONS

Many data models are implementing for traditional data analytics, and these are often static and limited to addressing rapidly changing and unstructured data. IoT devices produce billions of data samples, which are applied to find correlations between a large number of sensor inputs and a variety of external variables. Traditional data analysis requires a model based on previous data along with professional analysis to establish a connection between variables. The ML algorithm begins with the actual variables and then searches automatically for several predictor variables and their connections. Hence, when we know what we need but don't have the necessary input variables to make a valid decision, the ML algorithm is quite useful. Many ML algorithms have been used to learn from data samples that are important for attaining a goal. In manufacturing environments, ML's analytical abilities are extremely helpful to make a more efficient decision. Machine learning assists in forecasting when an IoT device quality improves. To analyses any data samples generated in real-time from IoT networks, enabling ML to learn typical behavior and detect anything out of the normal.

11.8.1 Logistic regressions

Logistic regression (LR) is a supervised learning algorithm and a famous classification algorithm used to calculate the probability of a target variable [24]. It is a linear classification algorithm that capable of performing multiclass or binary classification problems. Conventional ML algorithms are generally expressed as $y \in \{0,1\}$. The prediction probability of the given positive class given a d -dimensional input $X = [x_1, x_2, ...x_d] \in \mathbb{R}^d$ can be calculated as follows:

$$p(y = 1 | x; \theta) = \sigma(w^T x + b) = \sigma\left(\sum_{i=1}^{d} w_i x_i b\right) \qquad (11.1)$$

From the Equation (11.1), $\theta = \{w = [w_1, w_2, ...w_d] \in \mathbb{R}^d, b \in \mathbb{R}\}$ are the model parameters. $\sigma(.)$ is a probability interpretation activation function that squashes the linear output into the range between 0 and 1. The goal of the classification algorithm training is to discover appropriate values for the variables $\theta = \{w, b\}$. For cost or loss functions minimization, the classification algorithm began with some random initialization values $J(\theta)$. Log loss (cross-entropy) is used for the LR algorithm. The predicted probability for query data was applied to the input to make the classification decision.

The goal is to extend the binary LR to handle multiclass prediction problems, which can be expressed as $y \in \{1, 2, ..., c\}$. In the output layer, the softmax function is employed to normalize the output into probabilistic values. By reducing the loss function value, a collection of parameters c may be approximated such as $\theta = \{W = [w_1, w_2, ...w_c]^T, b = [b_1, ..., b_c]^T\}$. At the inference step, the class label will be allocated to the one with the highest classification probability using the trained model parameter. Anomaly attack detection [25], classifying IoT devices in the smart home [6], accident prediction [26], and detecting DDoS attacks [27] are some examples of IoT applications that can be solved by using the LR algorithm.

11.8.2 Decision tree

A decision tree (DT) is a tree-based supervised learning technique for solving regression and classification problems [28]. Nodes and edges form a decision tree. The node may be considered as just a feature, and the edges can be considered as

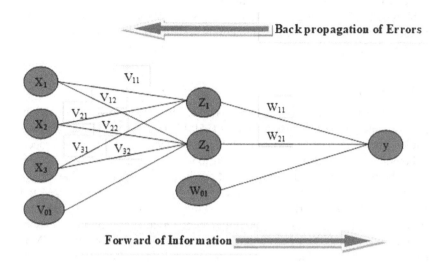

Figure 11.5 Architecture of BPNN algorithms.

just a classification constraint. DT's learning process involves employing a recursive approach to identify optimal features that can distinguish training data. The feature that will divide the entire population into two or more homogeneous sets based on the Gini impurity or information gain is chosen as the root node.

The DT algorithm's goal is to build a learning predictive model of the class or target value based on basic decision rules inferred from training data. The sub-population or sub-node will be chosen based on similarities (the same class is a high similarity), and this process will be repeated until all subsets are natural. The DT method requires that the input values be normalized to numerical values, and its tree structure allows for both nominal and numerical values to be used. However, the "curse of dimensionality" and an over-fitting problem impact the DT technique due to large input dimensionality. Figure 11.5 shows the DT algorithm and the DT algorithm can be used to solve a variety of IoT-related applications such as health care [28], clinical disease prediction [29], and intrusion detection [30].

11.8.3 Random forest

Random forest (RF) is a well-known machine learning ensemble technique that uses a large number of DTs as the basic classifier. Individual DTs are constructed using the random sampling feature, which also serves as a training example for diversity and improves the performance of various classifier systems. The random sampling method reduces the correlation between individual DTs and various prediction errors, while the aggregating function smooth's out huge prediction variances, resulting in a robust classification system with strong generalization capabilities. RF has high generalization ability and is used in a variety of IoT applications, including health care monitoring system [31], big data classification [32], smart building [33], and anomaly attack detection [34].

11.8.4 Support vector machine (SVM)

SVM is a supervised classification method that can be used to solve classification, regression, and outlier detection problems [35]. It's mostly used to address classification problems. The SVM algorithm's main objective is to discover a decision boundary (hyper-plane) in an N-dimensional space that categorizes the data points. The objective function of the SVM can be created to enhance the margin while minimizing certain instances. There are two types of SVM: linear SVM and nonlinear SVM. For linearly separable data, the linear SVM is used. The term "linear separable data" refers to the ability to divide a dataset into two classes using only one straight line. For nonlinearly separated data, the nonlinear SVM is used. The dataset is nonlinear, which means it can't be classified using a straight line. The SVM algorithm has been used in a variety of IoT applications, including smart city [36], IoT bot-net detection [37], lung cancer diagnosis [38], expiry prediction [39], and road signaling system [40].

11.8.5 Naïve Bayes

The naive Bayes (NB) algorithm is a supervised learning approach based on the Bayes theorem that is mostly used to solve classification problems [41]. NB is one of the most popular and simple classification methods. It helps with the development of a quick machine learning technique that makes precise forecasts. It is mostly utilized in text categorization with a large-scale training dataset. It is a probabilistic classifier, which means it predicts based on the probability of an object. NB is a probabilistic classification method because it predicts based on the probability of an object. The NB can be described as

$$p(y \mid x) = \frac{p(x \mid y)p(y)}{p(x)} \tag{11.2}$$

Here, $p(y \mid x)$ is represents a probability value, $p(y)$ is represent the prior, and $p(x)$ is represents evidence. In the classification process, labels are assigned based on posterior probabilities. The evidence, on the other hand, is a constant that can be destroyed out with ease. $p(y \mid x) \infty p(x \mid y)p(y)$ can be written into $p(y \mid x) = \prod_{j=1}^{d} p(x_j \mid y)p(y)$ for d-dimensional input x due to the hypothesis of independence features. Logarithms are often used in training to prevent the issue of floating-point underflow and the class label $\hat{y}$ can be assigned as follows,

$$\hat{y} = \arg\max_{y \in \{1,2,\dots,c\}} \ In\, p(y) + \sum_{j=1}^{d} p(x_j \mid y) \qquad (11.3)$$

Both discrete and continuous data can be modeled employing using the NB algorithm. The discrete probability can be calculated using frequency, and the continuous feature can be calculated using the density function. When dealing with data imbalanced problems, it takes into account the prior class distribution. Like DT, this also suffers from the "curse of dimensionality." Before solving a given problem, the feature extraction method can be applied to unstructured data in many cases. NB is an effective classification algorithm when the feature independence assumption is not considerably violated. The NB algorithm has been used to solve a variety of IoT applications, including food monitoring systems [42], water cleanliness monitoring systems [43], and health care [44].

11.8.6 K-nearest neighbor (KNN)

KNN is one of the most well-known supervised machine learning algorithms, instance-based nonlinear, and nonparametric classifier. When using the KNN algorithm, a majority vote is taken among the most similar samples of training data. Distance metrics, such as Minkowski distance, can be used to assess similarity which is described as follows:

$$D(x, y) = (\textstyle\sum_{j=1}^{d} |x_j - y_j|p)^{1/p} \qquad (11.4)$$

The KNN does not require any kind of training process and the distance calculation must be done for each data sample. Hence, this will result in a method that is less scalable for large datasets. In the same way, KNN is sensitive to outliers in the training set when choosing is small and makes KNN less discriminate when choosing is large. The KNN algorithm is widely used in IoT networks for a variety of purposes, including outlier detection [45], and fish farming [46].

11.8.7 K-means algorithm

K-means is a kind of unsupervised classification learning algorithm that is the most popular and well-known clustering algorithm due to its simplicity. This algorithm performs clustering tasks without class label information. The goal of the k-means algorithm is that the given data samples are segregated into values based on similarity. In the k-means algorithm, there are a variety of distance metrics available; the most commonly used is Euclidean distance, which has the following objective function:

$$\arg\min \sum_{i=1}^{N} \sum_{k=1}^{K} \gamma_{ik} \left\| x_i - \mu_k \right\|^2 \tag{11.5}$$

where μ_k represents the centroid of the kth cluster. γ_{ik} represents whether data samples fall into the kth cluster or not. The k-means algorithm is used to solve various problems in IoT environments such as multimedia [47] and energy [48].

11.8.8 Artificial neural networks

Artificial neural networks (ANNs) are a more advanced version of the linear classification algorithms that have already been discussed in previous sections. ANNs have an activation function that is nonlinear and allows them to model nonlinearity [49]. The relationship between input and output data samples is more difficult to see in ANNs because of an additional layer called the hidden layer. When compared with linear models, discovering the relationship between input and output data samples is more difficult with ANNs because they have a hidden layer. There are many applications that can be solved by using ANNs [50, 51].

11.8.8.1 Back propagation neural network (BPNN)

The BPNN is a supervised feedforward neural network that has three layers: input, hidden, and output [52] and Figure 11.6 shows its architecture. Each layer has adjustable weight and bias values, which serve as a connection between units whose activation values are always 1. The BPNN algorithm, which works in the direction of information, is fed forward and its error value is sent backward.

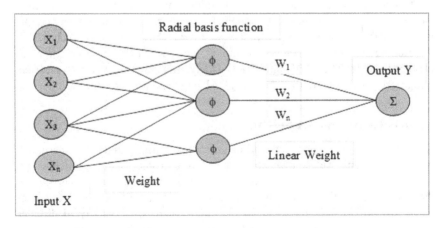

Figure 11.6 Architecture of RBF algorithm.

During the learning process, the weight is adjusted after every two passes. The BPNN performs two types of information flow: forward pass and backward pass. The information obtained from the input layer is then processed in a forward manner into the output layer via a hidden layer in the forward pass. When the networks are not achievable reasonable error for a given problem, the information is passed backward from the output layer to the input layer. The BPNN has four parts: initialization, feedforward, backpropagation error, and updating the weights. It is useful in a variety of fields, including pattern recognition, risk evaluation, and self-adaptive control, time-series analysis, and so on.

In the *initialize phase*, small random values are used to initialize the required parameters and connection weights. The hidden neurons are connected to the input layer neurons by their weights, which are represented by $[V_{ij}]$. The V is weight values of ith input neurons and j^{th} hidden neurons. The weight of hidden layer neurons is represented by $[W_{jk}]$, and they are connected to the output layer neurons. The value of bias is 1, which is represented by $[V_{0j}]$ and $[W_{0k}]$, and it improves the performance of the neural network. In the *feedforward phase*, input values are $X = (x_1, x_2,....,x_n)$ spread to each of the hidden units $Z = (z_1, z_2,....,z_k)$ after multiples with weight. Each hidden unit calculates the activation function and sends the result to the output units. The activation function is calculated in the output unit to form the net's response. Calculate the hidden neuron using the below formula:

$$z_{-inj} = V_{oj} + \sum_{i=1}^{n} x_i v_{ij} \qquad (11.6)$$

After applying the activation function, the following equation can be used to obtain the value of hidden neurons in the hidden layer:

$$Z_j = f(z_{inj}) \qquad (11.7)$$

The output units receive an output of the hidden units' values, and each output unit calculates its output values using weights as follows:

$$y_{-ink} = W_{ok} + \sum_{j=1}^{p} z_i w_{jk} \qquad (11.8)$$

Then, by applying the activation function to the values of the hidden neurons, the following result is obtained:

$$Y_k = f(y_{-ink}) \qquad (11.9)$$

In the backpropagation of errors, each output unit compares the results of its computed actual activation to the target, and the error signal is sent back to the input layer with adjusted weights and network error values. To meet the convergence conditions, the error backpropagation phase has been extended. In the BPNN, the error term plays a significant role, and the convergence is determined by the error rate. The user determines the value of the minimum error based on the solutions provided. This is how the error term is calculated:

$$\delta_k = (t_k - y_k) f(y_{-ink}) \tag{11.10}$$

Following the discovery of the error term, the values of each hidden unit are calculated using network error as follows:

$$Z_j = \sum_k^m \delta_j w_{jk} \tag{11.11}$$

Using the following formula, the error information is used to compute with the activation function.

$$\delta_j = \delta_{-inj} f(z_{-inj}) \tag{11.12}$$

Finally, in the updating of weight and bias phase, each output unit, like the input units, updates bias and weight, and then the weight tuning term is added to the old weight with the new calculated weight using the gradient descent-based learning rule. The delta learning rule and first-order derivate rule follow the gradient descent learning algorithm. The bias term and weight are initially calculated as follows:

$$\Delta w_{jk} = \alpha \delta_k z_j \tag{11.13}$$

$$\Delta w_{ok} = \alpha \delta_k \tag{11.14}$$

Finally, using the formula below, the new weight values are calculated:

$$\Delta w_{jk}(new) = \Delta w_{jk}(old) + \Delta w_{jk} \tag{11.15}$$

$$\Delta w_{ok}(new) = \Delta w_{ok}(new) + \Delta w_{ok} \tag{11.16}$$

Each hidden unit's bias and weights terms are updated, and the weight tuning term is given as

$$\Delta v_{ij} = \alpha \delta_j x_i \tag{11.17}$$

The tuning term for bias is given as

$$\Delta v_{oj} = \alpha \delta_j \tag{11.18}$$

Finally, the input and hidden layer weight values and bias term are calculated as follows:

$$\Delta v_{ij}(new) = \Delta v_{ij}(old) + v_{ij} \tag{11.19}$$

$$\Delta v_{oj}(new) = \Delta v_{oj}(old) + \Delta v_{oj} \tag{11.20}$$

11.8.8.2 Radial basis functions (RBF)

An RBF is an unsupervised learning method that considers a point's distance from the center and the architecture is shown in Figure 11.7. The RBF functions have two layers: first, the RBF algorithm is combined with its features in the inner layer, and then the outputs of the features are taken into account when determining the same output in the next time step, which is fundamentally a memory function [53]. RBF network performs a nonlinear mapping from input space R^n to the output space R^m. R^n is an input vector space that is denoted by x_i (for $i = 1, 2, 3....n$) and R^m is output vector space that is denoted by y (for $i = 1, 2....m$). The j-th hidden neuron of the RBF, which computes a Gaussian function as below:

$$Z_i(x) = \exp(-\frac{\|x - c_j\|}{2\sigma_i^2}) \quad j = 1, 2,....m \tag{11.21}$$

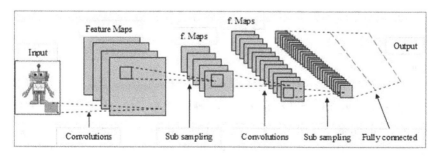

Figure 11.7 Architecture of CNN algorithms.

where, x represents an input feature vector with n-dimensional data samples. c_j represents the Gaussian vector center of i and σ_i represents the width of the hidden layer. Every hidden node center value is calculated by using the fuzzy C-means algorithm. The hidden layer width value σ_i is calculated by

$$\sigma_j = \sqrt{\frac{1}{m_j} \sum_{i=1}^{m_j} d^2(c_j - x_i)} \tag{11.22}$$

11.8.9 Deep learning methods

There are limitations to the ANN method in terms of generalization when dealing with unknown data samples. Then, to add more layers to ANNs, they are called deep learning methods. The deep learning method tends to be more accurate in terms of modeling [54].

11.8.9.1 Deep neural networks

Deep neural networks (DNN) can be defined as the simplest system of deep learning technique, which contains one input layer, multiple hidden layers for extracting more complex features, and one output layer [55]. DNN layers are arranged in a hierarchical order, with each layer having to serve as a function as the one before it. The backpropagation method can be used to learn the DNN method and perform end-to-end learning. The deep or fully connected layer is represented as follows:

$$h^{(l)} = g^l(W^{(l)T} h^{(l-1)} + b^{(l)}) \tag{11.23}$$

$W^{(l)T}$ and $b^{(l)}$ represent weight and bias values and g^2 represents activation function for the second layer. The total number of layers is represented as L, the output layer linearly transforms the previously hidden unit $h^{(L-1)}$ followed by the softmax function for probability scaling:

$$p(y \mid h^{(L-1)}) = \text{softmax}(W^{(L)T} h^{(L-1)} + b^{(L)}) \tag{11.24}$$

The weights and bias are the model parameters that needed to be calculated by reducing the loss function.

11.8.9.2 Convolutional neural networks

The convolutional neural network (CNN) algorithm, like the DNN, has multiple hidden layers. A CNN consists of an input, hidden, and an output layer and

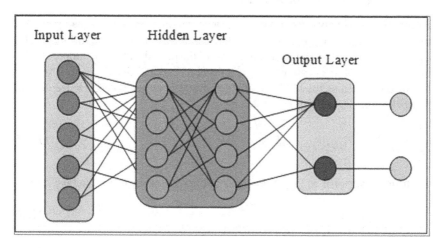

Figure 11.8 Architecture of RNN algorithm.

Figure 11.8 shows the architecture of CNN. The hidden layer has three types including convolution layer, pooling, and dense layer. In the convolutional layer, local samples can be extracted by learning numerous convolutional filters/kernels. Because of the weight-sharing scheme, CNN is effective at dealing with high-dimensional data. The selected feature maps were generated, and then a pooling layer can be used for downsampling after conducting the convolution operation [56].

The convolutional layer involves the input and passes the result to the next layer and is similar to a neuron's response to a specific stimulus in the visual cortex. Each convolutional neuron only processes data for the receptive field it is assigned to. Although fully connected FFNN can be used to study features and classify data, they are unfeasible for image classification. Due to the very huge input sizes associated with images, where each pixel is an applicable variable, it would need a very large number of neurons, even in a shallow architecture. To speed up the underlying computation, convolutional networks may include local or global pooling layers.

Small clusters, typically 2 × 2, are combined in local pooling. Global pooling affects all of the convolutional layer's neurons. Pooling can be separated into two kinds: average and maximum. The maximum value of each cluster of neurons at the previous layer is used in max pooling, whereas the average value is used in average pooling. Every neuron in one layer is connected to every neuron in another layer in fully connected layers. It works in the same way as a multilayer perceptron neural network (MLP). To classify the images, the flattened matrix passes through a fully connected layer. Each neuron in a fully connected layer receives input from all neurons in the previous layer.

11.8.9.3 Recurrent neural networks

Recurrent neural networks (RNNs) are deep neural networks that can recognize time-series data (sequential data) and their architecture is shown in Figure 11.9. The hidden layer of an RNN can feedforward the period and retain temporal information for subsequent inferences [57]. RNN has an additional set of parameters to calculate, such as the hidden-to-hidden transformation matrix. Hence, the full number of parameters can be written as:

$$\Phi = \left\{ W_{xh}, W_{hh}, W_{ho}, b_o, b_h \right\} \tag{11.25}$$

The input vector x_t at the ith time stamp, the feedforward pass can be written as

$$p(y \mid x_t) = \mathrm{softmax}(W_{ho}^T h_t + b_o) \tag{11.26}$$

$$\text{where } h_t = \tanh\left(\begin{bmatrix} W_{xh} \\ W_{hh} \end{bmatrix}^T [x_t, h_{t-1}] + b_h \right) \tag{11.27}$$

The present hidden unit h_t is computed based on the previously hidden unit h_{t-1} and present input units x_t. The prediction and the sample-wise feature can be learned at the same time because RNN is an end-to-end technique. To capture high-level contextual/temporal information, the deep RNN technique is necessary for learning the complex time-series data analysis. When it comes to solving time-series data, RNNs are particularly good at speech reorganization and machine translation.

11.8.9.4 Reinforcement learning

Reinforcement learning is a kind of deep learning method that is used to find the best policy for you to maximize a percentage of the total reward $J(\theta)$ parameterized by the reward $r_t = r(a_t, s_t)$ at each time t that is given when an agent takes an action $a_t \in A$ at state $s_t \in S$ [58]. Figure 11.9 shows the architecture of the reinforcement learning algorithm.

$$\arg\max J(\theta) = E_{t \sim p_\theta(t)} \left[\sum_t r(s_t, a_t) \right] \tag{11.28}$$

From the above equation, $P_\theta(t)$ is the interaction between the outer situation and RL agent. The outcome was based on two factors: agent policy and environmental dynamics. Agents can act following their policies $\pi_\theta(a_t \mid s_t)$. The environmental dynamic transition $s_t \times a_t \to s_{t+1}$ can be stated as follows:

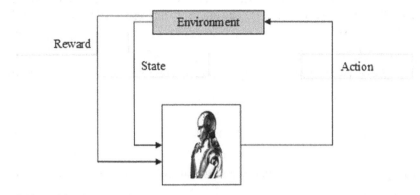

Figure 11.9 Architecture of reinforcement learning algorithm.

$P(s_{t+1} \mid s_t, a_t)$ and is generally unidentified. Overall, the entire trajectory $P_\theta(t)$ can be denoted as

$$P_\theta(s_1, a_1, .., s_T, a_T) = p(s_1) \prod_{t=1}^{T} \pi_{\theta(a_t, s_t)}(p(s_{t+1} \mid s_t, a_t)) \tag{11.29}$$

How to find the best trajectory is clearly stated in the equation above. If we are clever on the agent policy $\pi_{\theta(a_t, s_t)}$, or transition dynamic $(p(s_{t+1} \mid s_t, a_t)$, then we can determine the best approach P_θ. To maximize reward at each time step, the best trajectory is used. Model-free and model-based reinforcement learning methods are the two types of reinforcement learning.

(a) Model-free method

The model-free methods can learn a policy for deciding the best action to take specified a definite state. Based on how the policy is generated, it can be classified as policy-gradient and value-based [59]. The goodness of reaching a definite state or the goodness $V(s_t)$ of taking a certain action a_t given the state s_t is calculated using the *value-based* method, which involves learning a value function. The goodness function then calculates the total of future rewards from the present state s_t to the end s_T. Based on the estimated value function, the agent selects the action with the highest score. At each stage, the agent can choose the action with the highest score based on the estimated value function. It's a deterministic approach, so it might not be enough to solve a complex problem. Q-Learning, DQL, Dueling DQL, Double DQL, and Retrace are just a few examples of value-based reinforcement learning algorithms.

The policy-gradient methods offer an interesting paradigm by directly maximizing $J(\theta)$ with respect to the θ. The gradient with respect to the parameter θ can be derived as

$$\nabla_{\theta} J = E_{\theta}\left[\sum_{t} \nabla_{\theta} \log \pi_{\theta}(a_t \mid s_t)(R_t - b_t)\right] \tag{11.30}$$

$$R_t = \sum_{t'=t} \gamma^{t'=t} r(s_{t'}, a_{t'}) \tag{11.31}$$

where γ is the discount factor that focuses the agent's attention on recent rewards.

b_t is a baseline that is independent of the action or future state. Within a trajectory, the expected future return is sampled and aggregated. Actor-critic algorithms are related to the policy-gradient method when the predictable value function $V(s_t)$ is control in place of the constant baseline. This method has a lot less variance when it comes to the policy and value functions, which are referred to as actors and critics, respectively.

(b) Model-based method

The model-based technique varies from model-free techniques in that it is less concerned with the inner workings of the environment, and the rewards are approximated across sampling. A model-based technique, on the other hand, concentrates on the method to estimate the next state at each time step. Model-based reinforcement learning algorithms can produce good sample efficiency by learning transition dynamics of the environments directly. By learning the transition dynamics of the environments efficiently, a model-based reinforcement learning algorithm can produce the best sample efficiency. The sample trajectories are gathered and trained using supervised learning while performing the learning process [60].

11.9 CONCLUSIONS

IoT has received a lot of attention recently, and it's making life easier for people. It enabled various devices and objects to be connected over the internet. It allowed various internet-connected devices and objects to communicate with each other. The connected devices can collect meaningful data to enhance human effectiveness and competitiveness. On the other hand, machine learning is a technique of data analysis that automatically performs the creation of analytical models. It's a branch of AI-based on the knowledge that computers can learn from data, identify patterns, and make a decision with little or no human input. This chapter presented a discussion of the IoT application using a machine learning algorithm. There was also a graphical representation of the characteristics of IoT, applications, and various machine learning algorithms.

References

[1]   Kalla A, Prombage P and Liyanage M 2020 Introduction to IoT. *IoT Security: Advances in Authentication* 1–25.

[2]   Atlam H F and Wills G B 2019 *Advances in Computers*. Elsevier, pp. 1–39.

[3]   Tran C and Misra S 2019 The technical foundations of IoT. *IEEE Wireless Communications* **26**(3), 8. doi: 10.1109/MWC.2019.8752474.

[4]   Mahdavinejad M S, Rezvan M, Barekatain M, Adibi P, Barnaghi P and Sheth A P 2018 Machine learning for Internet of Things data analysis: A survey. *Digital Communications and Networks* **4**, 161–75.

[5]   Zantalis F, Koulouras G, Karabetsos S and Kandris D 2019 A review of machine learning and IoT in smart transportation. *Future Internet* **11**, 94.

[6]   Cvitić I, Peraković D, Periša M and Gupta B 2021 Ensemble machine learning approach for classification of IoT devices in smart home. *International Journal of Machine Learning and Cybernetics* 1–24.

[7]   Bharadwaj H K, Agarwal A, Chamola V, Lakkaniga N R, Hassija V, Guizani M and Sikdar B 2021 A Review on the role of machine learning in enabling IoT based healthcare applications. *IEEE Access* **9**, 38859–90

[8]   Lakhwani K, Gianey H, Agarwal N and Gupta S 2019 *Emerging Trends in Expert Applications and Security*. Springer, pp. 425–32.

[9]   Chiu D K, Xu T and Gondra I 2021 Random graph-based multiple instance learning for structured IoT smart city applications. *ACM Transactions on Internet Technology (TOIT)* **21**, 1–17.

[10] Abir S A A, Anwar A, Choi J and Kayes A 2021 IoT-enabled smart energy grid: applications and challenges. *IEEE Access* **9**, 50961–81.

[11] Agarwal V, Sharma S and Agarwal P 2021 *Computer Networks, Big Data and IoT*. Springer, pp. 709–16.

[12] Al-Turjman F and Baali I 2019 Machine learning for wearable IoT-based applications: A survey. *Transactions on Emerging Telecommunications Technologies* e3635.

[13] Li X, Lu Y, Fu X and Qi Y 2021 Building the Internet of Things platform for smart maternal healthcare services with wearable devices and cloud computing. *Future Generation Computer Systems* **118**, 282–96.

[14] Frank A G, Dalenogare L S and Ayala N F 2019 Industry 4.0 technologies: Implementation patterns in manufacturing companies *International Journal of Production Economics* **210**, 15–26.

[15] Alghanmi N, Alotaibi R and Buhari S M 2021 Machine learning approaches for anomaly detection in IoT: an overview and future research directions. *Wireless Personal Communications* 1–16.

[16] Latif S, Idrees Z, Huma Z and Ahmad J 2021 Blockchain technology for the industrial Internet of Things: A comprehensive survey on security challenges, architectures, applications, and future research directions. *Transactions on Emerging Telecommunications Technologies* e4337.

[17] Syam N and Kaul R 2021 *Machine Learning and Artificial Intelligence in Marketing and Sales*. Emerald Publishing Limited.

[18] Ullah Z, Al-Turjman F, Mostarda L and Gagliardi R 2020 Applications of artificial intelligence and machine learning in smart cities. *Computer Communications* **154**, 313–23.

[19] Cui L, Yang S, Chen F, Ming Z, Lu N and Qin J 2018 A survey on application of machine learning for Internet of Things. *International Journal of Machine Learning and Cybernetics* **9**, 1399–417.

[20] Celebi M E and Aydin K 2016 *Unsupervised Learning Algorithms*. Springer.

[21] Rathore S and Park J H 2018 Semi-supervised learning based distributed attack detection framework for IoT. *Applied Soft Computing* **72**, 79–89.

[22] Liang W, Huang W, Long J, Zhang K, Li K-C and Zhang D 2020 Deep reinforcement learning for resource protection and real-time detection in IoT environment. *IEEE Internet of Things Journal* **7**, 6392–401.

[23] Adi E, Anwar A, Baig Z and Zeadally S 2020 Machine learning and data analytics for the IoT. *Neural Computing and Applications* **32**, 16205–33.

[24] Hsieh F Y, Bloch D A and Larsen M D 1998 A simple method of sample size calculation for linear and logistic regression. *Statistics in Medicine* **17**, 1623–34.

[25] Hasan M, Islam M M, Zarif M I I and Hashem M 2019 Attack and anomaly detection in IoT sensors in IoT sites using machine learning approaches *Internet of Things* **7**, 100059.

[26] Mishra K N and Pandey S C 2021 Fraud prediction in smart societies using logistic regression and k-fold machine learning techniques. *Wireless Personal Communications* 1–27.

[27] Kumar P, Bagga H, Netam B S and Uduthalapally V 2021 SAD-IoT: Security analysis of DDoS attacks in IoT networks. *Wireless Personal Communications* 1–22.

[28] Manikandan R, Patan R, Gandomi A H, Sivanesan P and Kalyanaraman H 2020 Hash polynomial two factor decision tree using IoT for smart health care scheduling. *Expert Systems with Applications* **141**, 112924.

[29] Alabdulkarim A, Al-Rodhaan M, Ma T and Tian Y 2019 PPSDT: A novel privacy-preserving single decision tree algorithm for clinical decision-support systems using IoT devices *Sensors* **19**, 142.

[30] Taghavinejad S M, Taghavinejad M, Shahmiri L, Zavvar M and Zavvar M H 2020 Intrusion detection in IoT-based smart grid using hybrid decision tree. In: *2020 6th International Conference on Web Research (ICWR)*: IEEE, pp. 152–6.

[31] Kaur P, Kumar R and Kumar M 2019 A healthcare monitoring system using random forest and internet of things (IoT). *Multimedia Tools and Applications* **78**, 19905–16.

[32] Lakshmanaprabu S, Shankar K, Ilayaraja M, Nasir A W, Vijayakumar V and Chilamkurti N 2019 Random forest for big data classification in the internet of things using optimal features. *International Journal of Machine Learning and Cybernetics* **10**, 2609–18.

[33] Varma P S and Anand V 2021 Random forest learning based indoor localization as an IoT service for smart buildings. *Wireless Personal Communications* **117**, 3209–27.

[34] Tama B A and Rhee K-H 2018 An integration of pso-based feature selection and random forest for anomaly detection in iot network. In: *MATEC Web of Conferences*: EDP Sciences, p. 01053.

[35] Noble W S 2006 What is a support vector machine? *Nature Biotechnology* **24**, 1565–7.

[36] Shen M, Tang X, Zhu L, Du X and Guizani M 2019 Privacy-preserving support vector machine training over blockchain-based encrypted IoT data in smart cities *IEEE Internet of Things Journal* **6**, 7702–12.

[37] Al Shorman A, Faris H and Aljarah Iw 2020 Unsupervised intelligent system based on one class support vector machine and Grey Wolf optimization for IoT botnet detection. *Journal of Ambient Intelligence and Humanized Computing* **11**, 2809–25.

[38] Valluru D and Jeya I 2020 IoT with cloud based lung cancer diagnosis model using optimal support vector machine. *Health Care Management Science* **23**, 670–9.

[39] Nair K, Sekhani B, Shah K and Karamchandani S 2021 Expiry prediction and reducing food wastage using IoT and ML. *International Journal of Electrical and Computer Engineering Systems* **12**, 155–62.

[40] Sankaranarayanan S and Mookherji S 2021 *Research Anthology on Artificial Intelligence Applications in Security*. IGI Global, pp. 1003–30.

[41] Rish I 2001 An empirical study of the naive Bayes classifier. In: *IJCAI 2001 Workshop on Empirical Methods in Artificial Intelligence*, pp. 41–6.

[42] Ganjewar P D, Barani S, Wagh S J and Sonavane S S 2018 Food monitoring using adaptive Naïve Bayes prediction in IoT. In: *International Conference on Intelligent Systems Design and Applications*. Springer, pp. 424–34.

[43] Manaf K, Kaffah F, Mulyana E and Agnia N 2021 Implementation of Naïve Bayes algorithm in IoT-based water cleanliness monitoring system. In: *IOP Conference Series: Materials Science and Engineering*. IOP Publishing, p. 042007.

[44] Zhang Z and Zhang S 2021 Application of Internet of Things and naive Bayes in public health environmental management of government institutions in China. *Journal of Healthcare Engineering*. 2021, Article ID 9171756, https://doi.org/10.1155/2021/9171756

[45] Zhu R, Ji X, Yu D, Tan Z, Zhao L, Li J and Xia X 2020 KNN-based approximate outlier detection algorithm over IoT streaming data *IEEE Access* **8**, 42749–59.

[46] Islam M M, Uddin J, Kashem M A, Rabbi F and Hasnat M W 2020 Design and implementation of an IoT system for predicting aqua fisheries using arduino and KNN. In: *International Conference on Intelligent Human Computer Interaction*. Springer, pp. 108–18.

[47] Gupta M K and Chandra P 2021 Effects of similarity/distance metrics on k-means algorithm with respect to its applications in IoT and multimedia: a review. *Multimedia Tools and Applications* 1–26.

[48] Muthanna M S A, Wang P, Wei M, Rafiq A and Josbert N N 2021 Clustering optimization of LoRa networks for perturbed ultra-dense IoT networks. *Information* **12**, 76

[49] Sivanandam S and Deepa S 2006 *Introduction to Neural Networks Using Matlab 6.0*. Tata McGraw-Hill Education.

[50] Rathee G, Garg S, Kaddoum G, Wu Y, Jayakody D N K and Alamri A 2021 ANN assisted-IoT enabled COVID-19 patient monitoring. *IEEE Access* **9**, 42483–92.

[51] Kalaiselvi K, Velusamy K and Gomathi C 2018 Financial prediction using back propagation neural networks with opposition based learning. *Journal of Physics: Conference Series*. IOP Publishing, p. 012008

[52] Hecht-Nielsen R 1992 *Neural Networks for Perception*. Elsevier, pp. 65–93.

[53] Chen S, Cowan C F and Grant P M 1991 Orthogonal least squares learning algorithm for radial basis function networks. *IEEE Transactions on Neural Networks* **2**, 302–9.

[54] Goodfellow I, Bengio Y and Courville A 2016 *Deep Learning*. MIT Press.

[55] Miikkulainen R, Liang J, Meyerson E, Rawal A, Fink D, Francon O, Raju B, Shahrzad H, Navruzyan A and Duffy N 2019 *Artificial Intelligence in the Age of Neural Networks and Brain Computing*. Elsevier, pp. 293–312.

[56] Albawi S, Mohammed T A and Al-Zawi S 2017 Understanding of a convolutional neural network. In: *2017 International Conference on Engineering and Technology (ICET)*: IEEE, pp. 1–6.

[57] Medsker L R and Jain L 2001 Recurrent neural networks. *Design and Applications* **5**, 64–7.

[58] Sutton R S and Barto A G 2018 *Reinforcement Learning: An Introduction*. MIT Press.

[59] Dong Y, Shi Z, Chen K and Yao Z 2021 Self-learned suppression of roll oscillations based on model-free reinforcement learning. *Aerospace Science and Technology* 106850.

[60] Liu Q, Yu T, Bai Y and Jin C 2021 A sharp analysis of model-based reinforcement learning with self-play. In: *International Conference on Machine Learning*. PMLR, pp. 7001–10.

Chapter 12

Application of IoT in smart cities

Shubhangi P. Gurway

CONTENTS

DOI: 10.1201/9781003335801-12

12.1 INTRODUCTION

With the ever-increasing population and the massive migration of people to urban areas, management of all the facilities in a city and its governance is a quite challenging task for governments all over the world. To overcome this challenge, the concept of "smart city" came into existence. Smart city integrates the information and communication technology (ICT) tools to improve the operational efficiency, provide high-quality service to the citizens and help in smart governance. The main objective of the "smart city" concept is to optimize the services provided, promote economic growth and provide improved quality life to the citizens by using latest technologies. In order to achieve these objectives, the city must focus on how the best of available technologies can be used to add value to the standard of living and well-being of citizens.

A city's smartness is measured using some set of parameters such as the following:

- Technology-based smart infrastructure
- Environmental initiatives
- Effective and smart management of utility services
- Effective transportation
- Energy-efficient systems
- Improved standard of living of citizens.

But for the developing countries like India that entered into this "smart city" race, it's very important to focus on how the available technologies can be effectively used rather than how much technology is being used. So, IoT is one of the promising technological tools, even if simple yet effective as a source for managing all the things effectively. IoT simply connects all the things in the world through internet in order to collect the data from objects and manage them easily without any human intervention.

12.2 INTERNET OF THINGS (IOT) APPROACH TO SMART CITY

IoT basically applies the broadband network and uses internet as a conversion point to connect things to one another. IoT is engaged by the development of a few things and corresponding hardware. For things to be smartly connected to each other, they should have smart technology in them and these smart equipment are mobile phones and other related appliances that are going to be managed smartly. IoT-enabled smart cities give new technological lift to the citizens' life standards. IoT also needs millions of sensors to execute the interaction between the things being managed and controlled by IoT. Some low-power communication network standards are best suitable for interconnection between the devices. So, as per the required distance coverage and location of the devices to be connected, some available networks are presented:

1. Home area network (HAN) to monitor and control things at home by using short-range standard like Wi-Fi, Zigbee and Dash7.
2. Wide area networks (WAN), which are used for high-standard communication between the utilities, citizens and the service providers via broadband wireless like 3G and LTE.
3. Field area networks for interconnection between customers and substations (Hancke et al., 2012; Periasamy et al., 2021).

In IoT, sensing and processing of data is carried out by unified solutions, which involve Thingsspeak. It is an analytical platform developed by iOBridge, which aggregates, visualises and analyses the live data in the cloud. It is remarkable that for effective data storage and processing, cloud computing is very important (Talari et al., 2017; Sathiyamoorthi et al., 2021).

Some major IoT-related technologies which work together for the successful management and control of all the events in smart city processing and monitoring are described below.

12.2.1 Radiofrequency identification

One of the most commonly used devices in the IoT framework is radiofrequency identification (RFID), which generally uses a unique identification code to employ any of the technologies on concerned things, performs automatic identification and assigns a unique identity to it and establishes the network associated with digital information and service. Smart grids, health care and even smart parking in smart cities will use RFID as an effective tool to gather the data and make its maintenance easier.

12.2.2 Near-field communication

Near-field communication (NFC) is a set of communication protocols that are used for bidirectional communication of shorter distances (usually up to 4 cm). They are generally used for bootstrapping the simple setups. In smart cities, NFC can be used as a wallet, which helps us to use our phone as a bank card, public transportation card or access card.

12.2.3 LWPN like ZigBee

Zigbee is IEEE 802.15.4 standard-based low-cost wireless communication technology of LWPN, which transfers the data between the entities at home, hospital and traffic management systems.

12.2.4 Wireless Sensor Networks (WSNs)

WSNs have wide range of application in all areas of smart cities right from health care to tracking people's information like temperature tracking, blood pressure monitoring, their movement in smart home tracking and many more. WSN involve sensors, radio interface, analog to digital converter, memory and power supply for effective communication and interaction between the components in the system.

12.2.5 Dash7

Dash7 is low-power, long-range communication standard, which is used in low-latency applications like building automations. It generally operates at 433 MHz, which has better penetration and hence best suitable for smart meter development for utilities.

12.2.6 3G and long-term evolution (LTE)

3G and LTE communication standards are usually applied for wireless connection between mobile phones and data terminals. Now in IoT, LTE provides tremendous scope for establishing communication between the systems from home, business, energy, agriculture and advertising to health care. LTE provides built-in security to the systems and it is ideal mode for the transfer of data in a secure form. Similarly, LTE is more efficient than 2G and 3G and data transfer can be done at low cost per bit compared to 3G.

12.2.7 Middleware

Middleware generally aggregates the functionalities and communication abilities of all the interconnected devices in the system in the application layers.

With the combination of all the above IoT technologies, effective and smart governance of managing all the public utilities will bring the urban cities to a smart technological standard. In the next section of the chapter, the detailed applications of IoT along with effective combination of ICT tools are explained.

12.3 APPLICATION OF IOT IN SMART CITIES

12.3.1 IoT in smart energy management

Today's ever-demanding need for energy in urban areas poses a promising challenge to the government to how smartly they manage all the energy demand and consumption. Simplified energy management with cost savings is the need of an hour. Increasing energy prices, emission targets and operation cost are the reasons why one should monitor and control the energy consumption. However, different cities face massive challenges in this because of their diversity in infrastructure, geographical location and lack of real-time data reading and real-time actionable insights. Today the advancement in digital technology can not only help you to overcome the hurdles but with the power of analytics the administration can easily identify the potential areas of power consumption, the energy consumption rates, energy wastage and predict demand to achieve real savings (Ahmed et al., 2019; Kumar et al., 2021).

The application of IoT with artificial intelligence not only helps to control the energy consumption but also with the help of analytics, it gives you detailed insightful prediction of future energy demand from a particular area.

The use of IoT in smart energy management in city has the following components:

- Real-time database software
- Intelligent node for data analytics
- Network equipment and security expertise
- Standards for communication data models and services
- Performance and security testing

12.3.1.1 How does smart energy management work?

For efficient energy management, there should be connectivity between the energy user, energy providers and local governments to empower them with tools to remotely monitor, control, automate and share energy. Figure 12.1 depicts the application of IoT in managing the energy in smart cities. Details regarding the energy users like education institutions, hospitals, transportation, home and industries are sent via the cloud to the distributed database management system (DBMS). The DBMS will perform data analytics with data analytics tools like PowerBi to provide real-time insights from data from both the production and consumption sides. It also helps to find out the major power consumption areas and accordingly help the administration to take corrective actions (Santos and Ferreira 2019; Smart Energy Management).

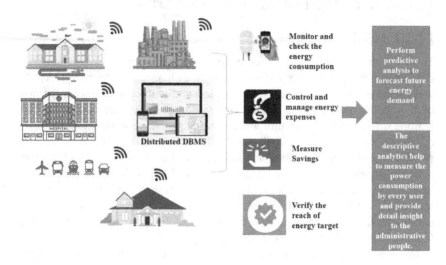

Figure 12.1 IoT-based smart energy management system in smart cities.

12.3.2 IoT in smart waste management

Solid waste produced by homes and buildings is called garbage. Every day, thousands of tonnes of waste is being generated in urban areas but as of now the waste collection systems in these areas are taken for granted by the citizens as well as respective municipal corporations. To overcome the issues like solid waste pollution, related health hazards and to maintain the standard of living, smart waste management system is the need of an hour.

As waste collection is among the essential facilities in a smart city, the use of smart technology to manage it is very necessary. So, IoT has the potential to greatly optimize collection services as well as reduce operational costs for cities (What is Smart Waste Management?).

12.3.2.1 How does smart waste management work?

An effective IoT-based smart solid waste management system is a combination of number of technological tools working, which is described below.

The smart waste management system consists of the following components:

- Smart bins
- Interfacing channel
- Cloud database
- Data analytics
- Mobile application

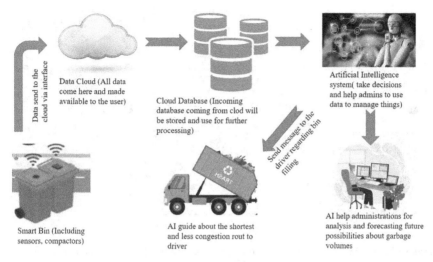

Figure 12.2 IoT-based smart waste management system.

A smart bin consists of a sensor and a compactor as a main functional unit. Once the garbage is disposed of into the bin, the ultrasonic sensor measures the level of waste inside it and the compactor operates to compress the garbage to the bin to its maximum capacity. Once the bin gets full, the sensor senses the maximum level and sends the signal to the cloud via interfacing channel like Wi-Fi/ 4G 5G network for further processing. The data coming from garbage bins from all the areas are collected in the DBMS, which perform real-time descriptive and predictive analysis with the help of artificial intelligence. The artificial intelligence system then sends the message to the drivers of respective areas whose bins get completely filled and triggers a signal. This process not only saves the fuel cost but also reduces traffic congestion issues due to unnecessary movement of garbage vehicles around the city. Along with this, the designed system helps the administrative personnel in the municipal corporation to predict the future waste generation in different areas to find out the highest waste generation parts of the city and manage them accordingly.

12.3.3 IoT in smart transportation

Managing transportation and traffic in an urban city is quite a challenging task. But with the application of IoT tools like smart sensors, advanced communication technologies and high-speed network, things become easier. With the emergence of smart technology and IoT, the world is entering the next stage of movement called as smart transportation (An Introduction to Smart Transportation: Benefits and Examples).

Figure 12.3 IoT-based smart traffic management system (An Introduction to Smart Transportation: Benefits and Examples).

The smart traffic management system is a combination of technological entities like sensors, camera, cellular routers and automation to monitor and control the traffic congestion as well as deployment of 5G technologies in vehicles to support the system.

12.3.3.1 How does smart traffic management system work?

Smart traffic management systems, which are included in the umbrella of "intelligent transportation systems" and sometimes called "intelligent traffic management", are automated systems that incorporate the latest advances in IoT technology.

The working of the system starts with the collection of data from all the areas through sensors and cameras, which help the person in the control room to monitor the traffic congestion in the city. Once congestion is detected, the intelligent adaptive control system is invoked, which causes dynamic adjustment to the system including traffic light, RAM signalling or rapid transit lanes. In order to avoid accidents, the connected vehicle technology adopts the technology based on the deployment of 5G networks into vehicles. Similarly, the system also gives priorities to emergencies, which also may be implemented with the sound sensors, which triggers a signal to clear the routes for the same (An Introduction to Smart Transportation: Benefits and Examples; Advanced Road Traffic Management System for Smart Cities).

12.3.4 IoT in smart health management system

Managing health is the prior concern for all of the city governments. It will be quite easier to handle any pandemic if digital tools are used to give personal assistance to patients, monitor their health remotely, for remote diagnosis and for tracking up-to-date health-related information of people. This can be possible by using smart technologies with IoT.

12.3.4.1 How does IoT-based smart health management work?

With the use of all the existing available resources, the health care sector can also manage all the things with IoT. For instance, during the COVID-19 pandemic, if the affected person is in serious medical emergency, the relatives can find the availability of bed and wards through a mobile application where database of all the hospitals are available through the tracking device installed in the bed itself. Once the patient is sent to the available bed and doctor prescribes the medication, the IoT-based medication tracking allows the doctor to monitor the impact of dosage he suggested on the patient's condition. Along with this, maintaining inventory of all medical equipment and medicines is an important concern for the hospital store department to smartly manage these things. The bed is equipped with RFID tags. Fixed RFID readers (e.g., on the walls) collect the info about the location of assets. Medical staff can view it using a mobile or Web application with a map. Similarly, RFID-enabled drug tracking helps pharmacies and hospitals verify the authenticity of medication packages and timely

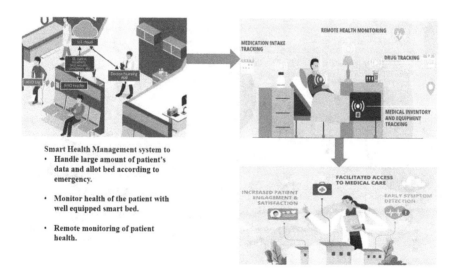

Figure 12.4 IoT-based smart health management system (Internet of Things (IoT) in Healthcare: Benefits and Challenges).

spot medication shortages (Internet of Things (IoT) in Healthcare: Benefits and Challenges). Once the patient gets discharged from hospital, with the help of smart monitoring device available with the patient, the doctors can easily monitor the status of health of the patient, which will get directly updated into the database of patient and avoid the revisit of patients to the hospital for minor issues. This will help hospital management team to ease the engagement of patients along with their health priorities.

12.3.5 IoT in smart surveillance

Safety, security and well-being of citizens is the first and foremost priority for any government. So, one should always think about this paradigm while thinking about smart city transformation. IoT along with ICT tools are now going to integrate every city right from highly accessible internet to sensors and cameras on road. So, IoT will be a promising tool in maintaining public safety.

12.3.5.1 How does IoT-based smart surveillance work?

IoT-enabled devices provide a significant amount of benefits to public safety like high network connectivity for ease of communication with related authorities for any security concern, real-time response and immediate call of action, receiving sensor-driven, secured and accurate data from citizens, which help to act and operate as per the measure of criticality and service optimization by highly assisted AI tools. For example, if there is an accident on the road, the traffic

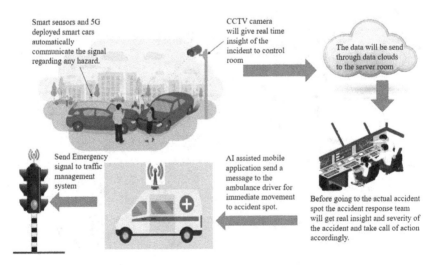

Figure 12.5 IoT-based smart safety management system (elaborate case of accident management).

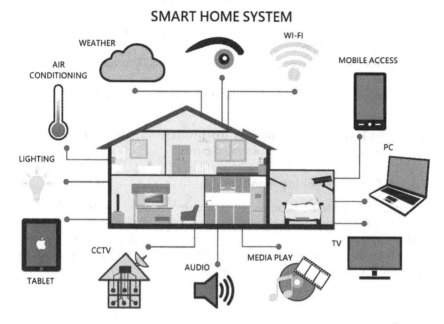

Figure 12.6 Application of IoT in a smart home (Visioforce Automation Hong Kong).

accident response team uses the data collected at their server room from CCTV cameras, speed sensors on vehicle for situational awareness and makes decisions before arriving at the accident spot. The AI assistance tool will help the traffic management team to clear the road for emergency transport.

12.3.6 IoT-enabled smart homes

Life becomes easier when all the stuff at home is done automatically without any human intervention. The IoT-based smart home management system makes this possible. With IoT, all the devices are connected to each other and to the remote monitoring and control units also. For instance, with the help of security camera in your home, you can easily see what is presently going on in your home by watching it through your phone. IoT-enabled smart homes are one of the best solutions connecting all the activities and to provide convenient, easy to command, safe and comprehensive management of things at home. With IoT, the devices are communicated to each other to such an extent that they can make individual decisions and take corrective action to provide all the smart service needed. One can easily manage their daily routines and smart home utilities at the point of their fingertip through IoT (Internet of Things (IoT) in Smart Home: Benefits and Uses; Visioforce Automation Hong Kong).

The key features of IoT-enabled smart homes are as follows:

- Humidity, temperature and oxygen level monitoring
- Visual management and remote activity access
- Closed-loop control system, which automatically controls the stuff until the required set of activity is not achieved
- Data collection for big data (Visioforce Automation Hong Kong).

12.3.7 IoT in miscellaneous applications for converting urban cities into smart ones

12.3.7.1 Smart air quality monitors

A smart city needs to be clean and pollution free. So, one important thing that needs to be monitored and controlled in a smart city is the quality of air. IoT-enabled smart air quality monitors not only monitor the indoor air quality but also find out the major geographical areas that are the main source of pollution and alert people for unsafe pollutant levels through flash messages in their smartphones if at home and sound an alert to the administration of the city.

12.3.8 Smart weather monitoring

The smart weather monitoring system monitors the weather condition, which can help the citizens and administration to take suitable steps. Similarly, a smart tracking system is connected to citizens' gadgets, which will give the updated status of the weather.

12.3.9 Smart water system

The smart water system with the help of various smart sensors will continuously measure the quality of water supplied to people and check the level of water in supply reservoir to facilitate flawless supply of water to the public domains. The smart sensors include chlorine residual sensor, total organic carbon (TOC) sensor, conductivity sensor and pH sensor that are used to monitor the quality of water. Similarly, a number of smart sensors are employed inside the pipelines to check the contamination concerns.

12.3.10 Smart parking system

Smart parking system enables the management of parking lots with ease and for effective traffic control. This system involves tracking of entry and exit of vehicles. In case of a crowded situation, the parking slot will be made available as per the priority for the cars. Every day, the departures and arrivals of vehicles can be monitored and mobile ticketing with the use of RFID can be issued with the use of a mobile application.

12.4 CHALLENGES FOR IMPLEMENTING IOT-BASED SMART CITIES

Security of data: For each and every facility being managed smarter, it is first and foremost important to make an interconnection or create a network between the systems. A huge amount of citizens' data is gathered and evaluated every day on an IoT platform, where there is a huge possibility of data being hacked or corrupted. Data vulnerability and cyber-attacks are one of the security concerns for public data. There are a number of interfaces through which data is unsecured, which are stored in clouds, high layer protocols and applications. For example, after installing any application in your phone, for example, WhatsApp, the application requires access permission of your phone. Doing so will share all the data of your phone, which can be the source of insecurity of your data.

Customization: Every city is different and different concepts have to be applied to get the city shifted from urban to smart one, but private IT companies who are the major players in the field want to use their single solution for all the situations.

Reliability of utility services: IoT-based systems are totally based on interconnection and communication between the components of the system. For any city to become smart, effective management of all the utility services right from power supply to broadband service is most crucial thing. We cannot treat a city as smart if it does not have 24 × 7 access to all the utilities as and when required. If there is a sudden failure in one of the communication channels or components in the system, the whole system fails and the system is then no more reliable.

Big data: Managing things by connecting around 50,000,000,000 devices and handling a huge amount of the data in the cloud is again a challenging task while applying IoT approach. This huge amount of data is transferred, stored and processed as well as analysed for the management of things. IoT substructures are one of the sources of big data (Talari et al., 2017).

Legal and social aspects: As huge amount of citizens' data is gathered and processed every day, and the IoT system works according to the provided data. The major players who likely manage all the services need to adhere to various local and international protocols and rules. Similarly, it is the responsibility of the concerned government to take care of and provide safe data interaction for smart governance.

Figure 12.7 Challenges for IoT-based smart cities.

12.5 FUTURE OF IOT-BASED SMART CITIES

IoT along with use of ICT tools will help in achieving the smart transformation of urban cities into "smart cities" throughout the world. With things becoming smarter, a number of municipalities would witness this smart transformation through enhanced traffic management, effective surveillance, increased energy efficiency with low power input and improved standard of living of citizens. People are getting familiar with the concept of applying smart technology to enhance their life standards. With lot of positive results in applying the concept using IoT, in future, almost all the parts of the world would get smarter and get the citizens to experience a new technologically advanced life.

12.6 CONCLUSION

With the ever-increasing population, growth in industrial sectors and rapid urbanization demands, smarter technologies are applied to manage all the resources associated with number of parameters like traffic control, reducing crime rates, waste management, managing energy consumption and water supply. The use of information and communication technology along with the emerging concept of IoT will be a promising solution for all the issues in helping the transformation of an urban city into a smart one, which is need of the hour. In future, IoT-based smart cities will not only help in effective governance but also boost the world's economy using new technologies.

Bibliography

Advanced Road Traffic Management System for Smart Cities (transportadvancement. com). Available at: www.transportadvancement.com/articles/advanced-road-traffic-management-system-for-smart-cities/

Ahmed, S. T., Sandhya, M. and Sankar, S. (2019). A Dynamic MooM Dataset Processing under TelMED Protocol Design for QoS Improvisation of Telemedicine Environment. *Journal of Medical Systems*, 43(8), 1–12.

An Introduction to Smart Transportation: Benefits and Examples. Digi International. Available at: www.digi.com/blog/post/introduction-to-smart-transportation-benefits

Hancke G., Silva, B. and Hancke G. Jr. (2012) The Role of Advanced Sensing in Smart Cities. *Sensors*, 13, 393–425.

Internet of Things (IoT) in Healthcare: Benefits and Challenges (scnsoft.com). Available at: www.scnsoft.com/healthcare/iot

Internet of Things (IoT) in Smart Home: Benefits and Uses. SCAND Blog. Available at: https://scand.com/company/blog/internet-of-things-in-smart-home/

Kumar, S. S., Ahmed, S. T., Vigneshwaran, P., Sandeep, H. and Singh, H. M. (2021). Two Phase Cluster Validation Approach Towards Measuring Cluster Quality in Unstructured and Structured Numerical Datasets. *Journal of Ambient Intelligence and Humanized Computing*, 12(7), 7581–7594.

Medagliani P., Leguay J., Duda A., et al. (2014). Internet of Things Applications – From Research and Innovation to Market Deployment. In Bringing IP to Low-Power Smart Objects: The Smart Parking Case in the CALIPSO Project: Delft, (The Netherlands): The River Publisher Series in Communication, pp. 287–313.

Periasamy, K., Periasamy, S., Velayutham, S., Zhang, Z., Ahmed, S. T. and Jayapalan, A. (2021). *A Proactive Model to Predict Osteoporosis: An Artificial Immune System Approach*. Expert Systems.

Sathiyamoorthi, V., Ilavarasi, A. K., Murugeswari, K., Ahmed, S. T., Devi, B. A. and Kalipindi, M. (2021). A Deep Convolutional Neural Network Based Computer Aided Diagnosis System for the Prediction of Alzheimer's Disease in MRI Images. *Measurement*, 171, 108838.

Santos, Diogo and João C. Ferreira (2019). IoT Power Monitoring System for Smart Environment. *Sustainability*, 11(19), 5355, https://doi.org/10.3390/su11195355

Smart Energy Management (capgemini.com). Available at: www.capgemini.com/fi-en/resources/smart-energy-management/

Talari, Saber, Miadreza Shafie-khah, Pierluigi Siano, et al. (2017). A Review of Smart Cities Based on the Internet of Things Concept. *Energies*, 10, 421.

Visioforce Automation Hong Kong – Smart Home, Home Automation. Available at: https://mindstec.com/in/

What is Smart Waste Management? (iotforall.com). Available at: www.iotforall.com/smart-waste-management#:~:text=Smart%20waste%20management%20soluti ons%20use,are%20ready%20to%20be%20emptied

Chapter 13

Novel vocal biomarker-based biosignal acquisition, analysis and diagnosis system

Archana Ratnaparkhi, Gauri Ghule, Pallavi Deshpande, Rohini Chavan, and Shraddha Habbu

CONTENTS

DOI: 10.1201/9781003335801-13

13.1 INTRODUCTION

Speech recognition (SR) is a technique that enables a computer to detect sounds produced in human speech. SR is invested in a variety of programs such as medical writing, marketing tools, the communications industry and learning resources. A few problems arise from the recognition phase of the unresolved issue despite significant progress being made in this field (1). It is a challenge to determine the appropriate divider for this particular problem. The conventional neural network classifier (NN) used in this project automatically adjusts their weights to improve their visual performance. Determining input parameters is a major challenge for NN. The reading level (LR) is one of the input components for a class divider. LR can slow down or speed up the process of learning by correcting weights in back-propagation. Lower LR trains the network less and may lead to overload. High LR makes the network deviate as it comes from local minima. LR should therefore be defined in advance and one cannot choose it naturally. This work introduces an effective way to use flexible reading quality to improve SR accuracy. Neural nets pose a serious drawback when it comes to locating local minima, convergence rate and instability in terms of network performance (2,3). As the literature survey indicates, gradient descent algorithms cause decrease in the errors such that the estimated output tends to match the required output to cause reduction in minimum squared error. Using a scalar multiplier with gradient causes expedited reduction in error. Learning rate parameter being the most crucial factor has been researched for decades to have its appropriate estimate. Too high LR might cause the network to get stuck in local minima while too low LR would cause an extremely slow network convergence (4). Adjustment of weights to achieve optimal performance is the key factor in designing neural nets (5). In Ref. (6), variable learning rate assignment is done to achieve high accuracy through assigning positive and negative signs to the partial deviates to minimize the error. Shiping Wen et al. (7) propose to use a fuzzy multilayer neural network to find dynamically changing weights and learning rates. Triangular membership functions and trapezoidal functions have been the most commonly used membership functions that suit various applications (8). Various adaptive learning rate evaluation methods have been studied for variable data sets and sizes. Most of these methods avoid manual calculation of the learning rate (9). It does not need to set the LR manually. The algorithm is tested for large scale data set and also comparable accuracy is achieved.

13.2 EXPERIMENTAL ANALYSIS

Figure 13.1 shows the SR classifier with three layers, namely input neurons, hidden layer neurons and output layer neurons. The number of neurons on the input side is decided by the classes of patterns in the data set. Usually, neural nets are accompanied by bias to achieve better fit of the data. The steps below describe the implementation of dynamic learning rate evaluation algorithm:

a. Consider a training set TT with output known.
b. Evaluate the forward propagation of TT through network.
c. Evaluate the weights and activation value at hidden layer using Equation (13.3).

For the hidden layer,

$$s_i = b_i V_i + \sum_{i,j} W_{i,j} + u_j \ where \ 1 \le i \le n, 1 \le j \le m \tag{13.1}$$

$$u_i = f\left(s_i\right) \tag{13.2}$$

$$f\left(s_i\right) = \frac{1}{1+e^{-s_i}} \tag{13.3}$$

d. Evaluate the weighted sum and activation value at output layer. For output layer,

$$s_i = b_i V_i + \sum_{i,j} W_{i,j} + u_j \ where \ 1 \le i \le k, 1 \le j \le n \tag{13.4}$$

e. Minimum squared error, which is a standard norm to find the error in the network, is evaluated and indicate by the given equation:

$$MSE = \frac{1}{2} \sum_i \left(C_i - u_i\right)^2 \tag{13.5}$$

Feedforward pass is completed at this point.

f. Further evaluations involve calculating error in backward direction at intermediate nodes. A derivative of sigmoidal function is used to evaluate error at every node is given as:

$$f\left(u_i\right) = u_i\left(1 - u_i\right) \tag{13.6}$$

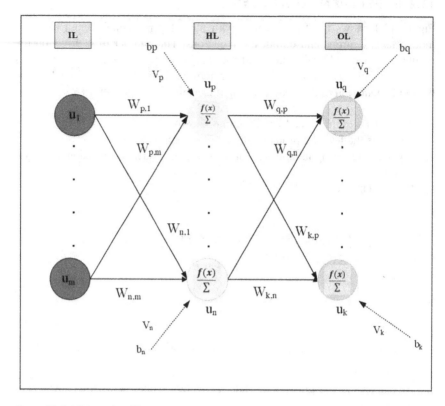

Figure 13.1 NN used in SR.

g. Evaluation of error at *i*-th output cell can be given as follows:

$$\delta_i = \left(\sum_{m:m>i} W_{m,j} \delta_j \right) f\left(u_i\right) \tag{13.7}$$

h. Evaluation of Error at each ith hidden cell can be shown as follows:

$$W_{i,j}^* = W_{i,j} + \rho \delta_i u_j \tag{13.8}$$

i. Using Equation (13.9), updation of weights in the hidden layer is done. The value of LR used for the first iteration is 0.1, which is responsible for controlling change occurring in each step.

$$LR\left(itr+1\right) = LR\left(itr\right) - \Delta LR \tag{13.9}$$

j.   The feedforward pass is run again by repeating steps 2 to 6 and MSE is calculated to verify the reduction in error in the output. It completes the first iteration of the NN.

k.   Dynamic LR is used to train the network by changing its value for every iteration. The change is determined by the MSE generated in the previous iterations. Equation (13.10) gives the computation of dynamic LR for every successive iteration 'itr+1'. LR represents the desired change in LR for every successive iteration. The assumption here is rate of change of LR is proportional to rate of change of MSE. Equation (13.10) represents LR from rate of change of MSE for previous two iterations where itr is current iteration and 'itr-1' is iteration prior to 'itr'.

$$\Delta LR = \frac{LR(itr)\big(MSE(itr-1)-MSE(itr)\big)}{MSE(itr)} \quad\quad (13.10)$$

l.   The dynamic LR varies between the values ranging from 0.1 to 0.000001. This approach allows the network to converge fast improving the training accuracy.

13.3 RESULTS AND DISCUSSION

The speech datasets used for experimentation are listed in Table 13.1. Figure 13.2 shows the plot of training accuracy of NN for different values of LR. Learning rate evaluation and optimal value generation cause the training process to take place smoothly without network getting stuck in the local minima. However higher value of learning rate, typically beyond 0.1 causes network to become unstable and keeps fluctuating up and down to cause an unstable output. Furthermore, underfitting is seen when learning rate is kept too slow, namely LR = 0.0005. Experimentation using TIDIGITS dataset involves evaluations done at learning rate equal to 0.05 to provide an accuracy of around 99 percent. However such high

Table 13.1 Speech data sets

Parameter	SMND	TIDIGITS
Number of samples	10000	4554
Sampling frequency	16 kHz	8 kHz
Number of speakers	100	207
Vocabulary size	Marathi language numerals "Shunya" to "Nau"	English language numerals 0 to 9, zero
Output classes	10	11
Utterances per class	10	2
Training samples	6250	2068
Testing samples	3750	2486

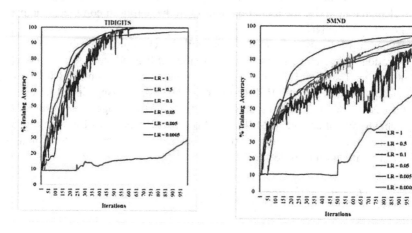

Figure 13.2 Classifier performance: Analysis of learning rates.

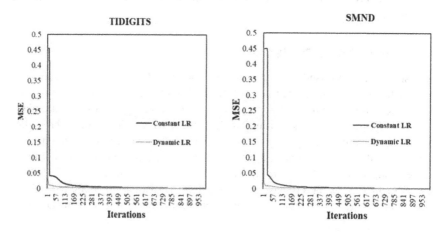

Figure 13.3 Performance training curves: Learning rates (constant and dynamic).

value is not achieved at LR = 0.1 .To have the same accuracy, number of iterations are high .With further reduction on learning rate, the number of iterations further increases to achieve high accuracy. Thus experimentation done on both the data sets strongly indicates that as single value of learning rate shall not be enough to have optimal accuracy. Instead dynamically changing LR is the solution. Figure 13.3 indicates this aspect. Model that involves constant LR clearly indicate from Figure 13.4 its inability to achieve the required high accuracy. A comparative performance of state-of-the-art techniques and proposed technique are listed in Table 13.2. The proposed technique with dynamic LR presents a maximum testing

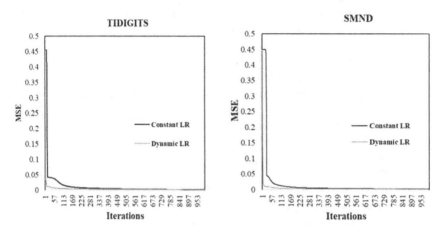

Figure 13.4 Significant convergence graphs.

Table 13.2 Comparative performance of state-of-the-art techniques and proposed technique

Sr. No.	Reference Classifier	Dataset	Testing accuracy
1	(Swami and Mulani, 2017) NN	TIDIGITS	94.67%
2	(Mulani and Mane, 2019) NN	TIDIGITS	95%
3	(Jadhav et al., 2021) NN	TIDIGITS	97%
4	(Deshpande et al., 2019) Dynamic multilayer perceptron	TIDIGITS	96.94%
5	Proposed approach NN with dynamic LR	TIDIGITS	99.03%

accuracy of 99.03 percent for TIDIGITS data set as compared to another state-of-the-art approaches.

13.4 USE OF BIOMEDICAL SIGNALS FOR AUTHENTICATION

Biological signals are records of the location, time, or duration of a biological event, such as a heartbeat or muscle contraction. The electrical, chemical and mechanical activities that take place during this biological phenomenon generally produce measurable and analytical signals. The most essential biomedical signals are divided into two types such as action force/ potential and event-related force. Electromyogram (EMG), electroneurogram (ENG), electrocardiogram (ECG) and electroencephalogram (EEG) are the existing action potential. An event-related energy (ERP) is a balanced brain response that is a direct result of a specific sensory, mental or motor event. Officially, it is any concentrated electro-physiological response to a stimulus. Our body produces various physical signals. The accessibility to these signals is important because these signals:

- May be internal (blood pressure)
- Can emit from the body (infrared radiation)
- Probably derived from tissue samples (blood or tissue biopsy)

All physiological signals can be grouped as:

- Biopotential
- Pressure
- Flow
- Dimensions (imaging)
- Displacement (velocity, force, acceleration)
- Impedance
- Temperature
- Chemical concentration and composition

The purpose of biomedical signal processing is to gather important information from the biomedical signal. Signal processing techniques can also be used to improve transmission, storage capacity and subjective quality and to emphasize or identify components that are interested in the measured signal. The transducer converts the physical signal into an electrical output. The transducer should respond only to the target form of the energy contained in the physical signal and exclude all other energies and should not be invasive.

13.5 BENEFITS OF USING BIOELECTRICAL SIGNALS AS BIOMETRICS

The use of bioelectrical signals provides a new approach to user authentication that includes all the important features of the previous traditional authentication. Among the most important reasons for implementing electrical bio-signals in user authentication are their scalability, uniqueness, universality and resistance to spoofing, while other traditional biometrics such as face size, hand size, fingerprint and voice can be spoofed. Biometrics aims to facilitate the identification management system to achieve a high level of accuracy through the use of unique and measurable physical and behavioural characteristics of individuals. The limitations of using traditional biometrics are that they contain unique identifiers, but they are not confidential or confidential to any individual. For example, individuals leave their fingerprints on everything they touch, iris patterns appear anywhere, faces appear and voices are recorded. The presence of biometric prints in public allows intruders to take these prints and copy them as originals, thereby deceiving the system. The use of bioelectrical signals as biometrics provides a number of benefits to identity management systems. In addition to their specificity, bioelectrical signals are confidential and secure to an individual. They are hard to imitate. Therefore, a person's identity will not be duplicated, thereby protecting the privacy and confidentiality of users. An individual's biological information is genetically regulated by deoxyribonucleic acid (DNA) or ribonucleic acid (RNA) proteins. After all, proteins are responsible for the specific existence of certain parts of the body. Similarly, organs such as the heart and brain are made up of protein tissues called myocardium and glial cells, respectively. Therefore, electrical signals emitted from these organs show specificity in individuals. Bioelectrical signals such as electrocardiogram (ECG) and electroencephalogram (EEG) can be used for biometric applications. Studies show that heart rhythms and stimuli of the brain's electrical activity are recorded on the ECG and EEG, respectively. Individuals have unique characteristics, so it may be advisable to use them as biometrics for identity verification. Convenient features for using ECG or EEG signals as biometrics include universality, scalability, specificity and robustness. In addition, they have the inherent feature of vitality that characterizes life signs that provide strong protection from spoof attacks. Unlike traditional biometrics, the ECG or EEG is highly confidential and secure to an individual, making it difficult to duplicate. In addition to the EEG and ECG, a surface electromyogram (SEMG), an electrical expression of muscle movements, can be used as a biometric feature. Accurate signal recognition from EMG offers a unique advantage over two traditional electrical biosignal paths, the electrocardiogram (ECG), and the electroencephalogram (EEG), which allows users to customize their own signal code. Surface electromyogram signals are commonly used as source signals in hand and wrist gesture recognition. With four to six electrodes mounted on the upper limb, the classification accuracy of the 10 signals reaches greater than 95 percent under the pattern recognition framework. Another new technique is being adopted these days for human identification based on eye blinking waveform extracted from 'electro-oculogram' signals. The eye blink signal is extracted and applied for identification and verification functions. The preprocessing phase involves the decomposition of the empirical mode to separate the electro-oculogram signal from the brain waves. Then, a timeline of the eye blink waveform is used for feature extraction. Finally, linear discrimination analysis was adopted for classification. The researchers claimed the best accuracy of 97.3 percent and so far have confirmed them with an equal error rate of 3.7 percent. The results obtained confirm that the eye blink wave contains discriminatory information and therefore it serves as the basis for human recognition work. Figure 13.5 shows biometric system's workflow. The enrolment stage includes preprocessing, feature extraction and template storage. Testing stage includes preprocessing, feature extraction and matching.

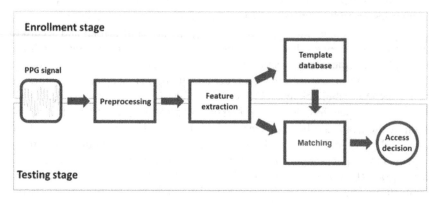

Figure 13.5 Framework: Biometric system.

13.5.1 Use of biomedical signals in multimodal biometric systems for high-security applications

The use of biomedical signals in multimodal biomedical signals such as EEG, ECG, SEMG and electro-oculogram can be used for the identification and verification for a high level of security, among other features in multimodal systems. For example, a novel multi-biometric system formed as a result of a combination of an ordinary biometric face and an electrocardiogram (ECG) of another biometric fingerprint is considered the least obstacle to effective personal authentication. Some researchers have even implemented multi-model authentication using a combination of ECG and fingerprint score levels and quoted an accuracy of up to 90 percent at rest. The future of biomedical signal analysis is very promising. We can expect innovative health care solutions that improve the quality of life of all.

13.6 RECENT ADVANCES AND APPLICATIONS IN MEDICAL IMAGING TECHNIQUES

A rapid growth in medical science and the invention in various medicines will benefit the whole civilization. Current science and technology advancements also help in doing wonders in the surgical field. But the proper and accurate diagnosis of diseases is the prime important task before the treatment using X-ray images as seen in Figure 13.6.

Better diagnosis is possible with the help of most advanced and sophisticated bio-instruments. The medical images play a very important role in the diagnosis of various diseases, research, teaching and therapy offered by the doctors. The thought process behind medical imaging is to represent anatomical structures of the body using X-ray and magnetic resonance imaging (MRI). Instead of anatomy, physiologic function is more useful. Medical imaging has greatly influenced the medical field due to tremendous growth in computer and image technology. The

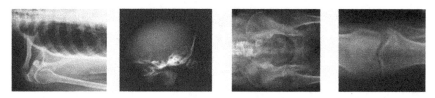

Figure 13.6 Significant X-ray images.

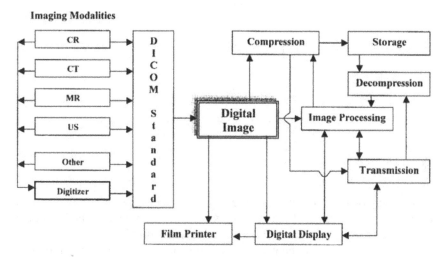

Figure 13.7 Overview: Digital Imagery module.

accurate diagnosis of any disease is possible using good quality medical images. It is required to understand some convenient processes in detail to store and retrieve images for future use. A conceptual framework of a digital image within a medical digital imaging system can be seen in Figure 13.7. Different medical imaging sources will be used to get a look inside the patient. These sources depend on how signal is travelling right through a patient. These signals interact with the tissues of the patient. The signals coming out of the body are detected and inside image of the patient is to be taken. Further, diagnosis can be done by processing these images.

13.6.1 Applications advances in medical imaging techniques

13.6.1.1 X-ray radiography

Radiography is one of the very popular diagnostic techniques that uses ionizing electromagnet radiation like X-ray to view the objects. X-ray is a high-energy

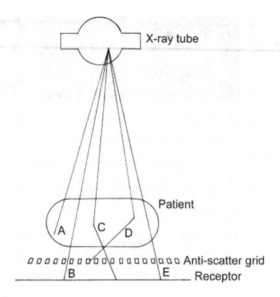

Figure 13.8 Profile: X-ray.

electromagnetic radiation. It penetrates solids and ionize gas. X-ray operates on a wavelength between 0.01 and 10 nm. X-ray passes through the body and the signals are absorbed or attenuated at differing levels. A profile is created according to the density and atomic number of the different tissues. The X-ray profile is registered on a detector creating an image as shown in Figure 13.8. Different images are taken from different parts of the body. The scientist needs to observe the image very carefully and identify the type of imaging from the patient. Over the years, medical imaging plays significant role in the early detection, diagnosis and treatment on many diseases. In some cases, medical imaging is the first step in preventing the spread of cancer through early detection and, in many cases, it is possible to cure or eliminate the cancer. CT scan, MRI, ultrasound and X-ray imaging are very important tools in the fight against several diseases. Medical imaging is also used to build accurate computer models of the body systems, organs, tissues and cells, which are used in teaching anatomy and physiology in the medical schools. The risks of X-ray radiography are exposure to ionizing radiation. This increases the possibility of developing cancer later in life.

Tissue effects such as cataracts, skin reddening, and hair loss occur at relatively high levels of radiation exposure.

13.6.1.2 X-ray computed tomography

Computed tomography (CT) is a diagnostic technology that combines X-ray equipment with a computer as seen in Figures 13.9 and 13.10. Cathode ray tube

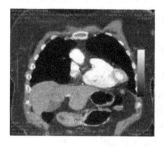
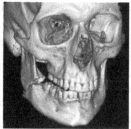
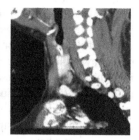

Figure 13.9 CT scan applications.

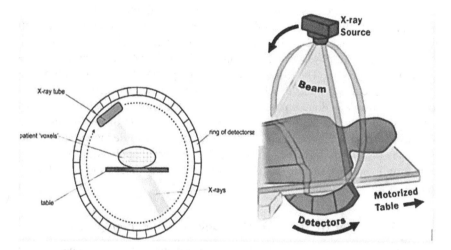

Figure 13.10 X-ray system.

display produces a cross-section of the images of the human body. The radiographic film is replaced by a detector, which measures the X-ray profile. Inside the CT scanner, a rotating frame, which houses the X-ray tube is mounted on one side and the detector situated on the other side. A beam of X-ray is generated as a rotating frame spins the X-ray tube and detector around the patient as shown in Figure 13.9.

Medical imaging refers to the techniques and processes used to create images of the human body for various clinical purposes such as medical procedures, diagnosis and medical science including the study of normal anatomy and function. It is a part of biological imaging and incorporates radiology, endoscope, thermograph, medical photography and microscopy. Measurement and recording techniques such as electroencephalography (EEG) and magnetoencephalography (MEG) are not primarily premeditated to construct images but which produce data disposed to be represented as maps and they can be seen as forms of medical imaging. The mathematical sciences were used in a broad way for image processing. They had little importance in biomedical work until the development of computed tomography (CT) for the imaging of X-rays and isotope production tomography. MRI (magnetic resonance imaging) ruled over the other modalities in many ways as the most revealing medical imaging methodology. These are all well predictable techniques, computer-based methods are being explored in the application of ultrasound and electroencephalography. Also it is used in the modern techniques of optical imaging, impedance tomography and magnetic source imaging. Although the final images obtained from many methods have similarities, but the technologies used and the parameters represented in the images are very diverse in characteristics as well as in medical effectiveness. Even unusual mathematical and statistical models have been used. Several other techniques have been investigated to enable CT, MRI and ultrasound scanning software to produce 3D images for the physician. Traditionally CT and MRI scans produced 2D static output on film. Then to create 3D imagery, many scans are completed and a 3D model is produced, which can be manipulated by medical doctor. If a detailed survey is done on medical imaging, the work will definitely afford a tangible outline on the past, present and future aspects of this technology. Visualization equipment and graphics workplace has afforded many dissimilar processes and ways of medical imaging. Out of these, application of wavelet transform in medical images, segmentation of medical images and virtual medical imaging subsystems are of supreme importance.

13.6.1.3 Wavelet transform Technology in medical imaging

The wavelet technology is widely applied to the domain of medical imaging and wavelet transform and inverse transform algorithms are introduced. Wavelet technology has been used in ECG signal processing, medical image compression, medical image reinforcing and edge detection, and medical image register. The image can be expanded in terms of the 2D wavelets. At each stage of the transform, the image is decomposed into four-quarter size images. For N-by-N image, the image is decomposed into four $N/2$-by-$N/2$ images for that stage of the transform. Applications of wavelet transform in medical image processing is used for image compression. To meet the demand for high-speed transmission of the image in efficient image storage and remote treatment, efficient image compression is essential. In recent times, some novel and very promising methods have emerged in the image compression algorithm based on wavelet transform, such as wavelet packet transform. According to the properties of its multi-scale, direction and local characteristic, determining the local maxima of wavelet coefficients provided the image edge features. Medical image enhancement and edge detection are very important methods in breast imaging.

13.7 APPLICATIONS OF WAVELET TRANSFORM IN MEDICAL IMAGE PROCESSING

13.7.1 Segmentation of medical images for best diagnosis

Computer vision typically identifies three processing steps before object recognition like image enhancement, feature extraction and grouping of similar features. In the image segmentation process, the pixels are grouped into regions based on image features. The objective is to divide an image into various pixel regions that together represent objects in the scene. A recently proposed oscillator network called the locally excitatory globally inhibitory oscillator network (LEGION). It is possible to achieve fast synchrony with local excitation and de-synchrony with global inhibition, which makes it an efficient computational structure for grouping like features and identifying dissimilar features from an image.

13.7.2 LEGION model

LEGION was proposed by Terman and Wang as a geologically reasonable computational structure for picture investigation and has been used fruitfully to fragment binary and grey-level images. It is a group of recreation oscillators, everything is constructed from an excitatory unit and an inhibitory unit. 2-D network architecture with four-neighbourhood coupling is also used at this point.

13.7.3 Segmentation algorithm

There are three instinctive criteria for defining groups on an image. The first is that the influential image should be generated from both consistent and brighter parts of the image. Second, brighter pixels should be considered analogous to wider ranges of pixels than darker ones. The third criterion specifies that the limits of sections are given where pixel intensities have relatively large changes. Three-dimensional segmentation is readily obtained by using 3-D district kernels. The segmentation algorithm has been applied on 2-D, CT and MRI medical datasets of the human head. The customer gives six input parameters like the probable district, the recruiting district, the threshold, the authority of the adaptive-tolerance-mapping function and the acceptance range-variables. With LEGION network, along with its biological plausibility, it is mainly possible to do a parallel-hardware implementation, which will be significant for real-time segmentation of volume data sets. Physical segmentation generally provides the most excellent and most consistent results when identifying structures for a particular medical job. One goal of medical image segmentation is to divide white substance and grey substance. This algorithm is proposed to be more supple for segmenting a multiplicity of structures. However, it is interesting to note that once the brain is segmented, using this algorithm, we can perform the partition of grey substance from white substance. Layers of LEGION networks are valuable computational structures that are capable of combining and segregating depends on fractional results from preceding layers and thus may further improve segmentation presentation. The network architecture is agreeable to VLSI chip execution, which would build LEGION, a viable architecture for real-time segmentation.

13.7.4 Superimposing a medical image within the subject

For superimposition of the medical image, within the subject itself, a virtual medical imaging system has been used in which three-dimensional (3D), stereoscopic, motion (if necessary) medical image is superimposed within the subject using a transparent head-mounted demonstrate. For displaying more logically, the 2D images are restructured as 3D images using computer graphics. For stereographic sights, a head-mounted display giving individual images for each eye is functional. Fundamental reality has been accepted in representing information collected from medical imaging devices like echography, MRI and CT. The medical imaging system consists of three subsystems: statistical information, input and output, and 3D object generations and image merging subsystem. The imaginative echo images in the 3D object production subsystem is obtained by measuring 28 consecutive B-mode echo images of the heart using a sector scanning trans-oesophageal probe. When the virtual imaging system was applied to echocardiogram, the virtual heart image changed its size, orientation and binocular parallax approximate for both the viewer and the subject matter. The echo image was just pasted one onto the subject. It was somewhat improved in this system due to the stereographic images. Stereographic image certainly has depth. This virtual medical imaging system will be valuable to health centres for purposes such as surgical planning in the future.

13.7.5 Medical image repository and image categorization

For a decade now, medical imaging has become a vital component in the medical field. However, the progress of the Internet has made possible for medical images that are available in large numbers in online repositories and other health-related sources. These images symbolize a precious resource for later use and are of prime importance for medical information recovery. Medical image depository plays a significant role in the hospital workflow. The operations in the hospital workflow like image storage and retrieval, screening and post-processing can be saved as a generic repository component. With the support of these operations, diverse modalities can build up their own medical application.

13.7.5.1 Medical image categorization using a texture-based symbolic description

In the field of medical image indexation, automatic classification provides the means for extracting some useful information from medical images. The compressed symbolic image representation conveys sufficient original texture information to be acquired far above the ground recognition rates, in spite of the compound background of multimodal medical image classification. Medical image classification construction, based on a latest type of image descriptor, aims to precisely take out the modality like MRI, X-ray and the anatomical region present in the medical images .When publishing using Internet, the images are suffering further transformations such as resizing, harvesting, high compression, superposed drawings and explanations. Thus the unpredictability of the images have increased. The physically powerful inter-class similarity between some classes improved the complexity level in categorization. The categorization approach consists of three phases: the removal of statistical and texture image-feature sets to explain the image visual content. Every image is denoted by a vector of 16 blocks. The features will be extracted from each block to describe its content. A k-nearest neighbour classifier is utilized, with the help of the first (1NN), the first three (3NN) and the first five (5NN) neighbours. For computing distances between nominal representations, value difference metric (VDM) is used to estimate the likeness between two symbolic features. The suggested feature representation or transformation technique is very similar to vector quantification (VQ), where the blocks of pixels are tagged with the indexes of the prototype blocks. But this approach is more receptive to image rotations and translations. Therefore it is rarely used. However, the recent digital acquisition equipment are following standard acquisition measures, and the images rarely present major variations to rotation and translation.

13.7.6 Web-based interactive applications of high-resolution 3D medicinal image data

The demands for sharing medical image data on the Internet for computerized visualization and analysis are growing consistently such large memory required data size, ranging from several hundreds of megabytes to several dozens of gigabytes, severely placing a strain on storage systems and networks, and it may hinder the expansion of Web-based interactive applications. Because of the limitations of the available Internet band width, Web-based interactive applications are limited to low- or medium-resolution image data. These kinds of data sets are often inadequate for consistent use in clinical analysis. Distinctively, first partition the 3D data into buckets and then squeeze each bucket discretely. Also an indexing construction for these buckets has to be done to professionally support typical queries such as 3D slicer and region of interest (ROI). Thus, only the related buckets are transmitted instead of the whole high-resolution 3D medical image data.

13.7.7 Data access optimization

To decrease the dependence on server in terms of disk access and communication costs, we use two methods. One is incremental transmission. A customer can send some bucket IDs along with a doubt to inform the server that those buckets are available locally and are not required to retransmit again. In multi-user surroundings, instead of recovering data buckets for each of the queries separately, allow them to split disk access by retrieving the data buckets inside the MBB, which encircles all the ROIs in one chronological admittance.

13.8 CONCLUSION

In this work, rigorous experimentation has carried out for the technique proposed in this work to develop efficient SR systems. The proposed classifier adapts the LR dynamically over iterations while maintaining training accuracy. The results clarify how higher LR results in unstable training and lower rates fail to train. Dynamic LR can accelerate training and alleviate the pressure of choosing an LR. The recognition accuracy increases to 99.03 percent using the proposed technique. Biomedical imaging has seen truly exciting advances in recent years. Newly invented imaging methods can now reflect internal anatomy and dynamic body functions heretofore only derived from textbook pictures, and their applications to a wide range of diagnostic and therapeutic procedures can be possible. Not only improvement in computer technology, but development will require continued research in physics and the mathematical sciences (e.g. artificial intelligence), fields that have contributed greatly to biomedical imaging and will keep continuing to do so. The major topics of recent interest in the area of functional imaging involve the use of MRI and positron emission tomography (PET) to explore the activity of the brain when it is challenged with sensory stimulation or mental processing tasks. The emerging imaging methods have the potential to address major medical and societal problems, including the mental disorders of depression, schizophrenia, and Alzheimer's disease and metabolic disorders such as osteoporosis and atherosclerosis. Although computing speed certainly has reached the point where iterative methods are clinically feasible for 2D problems, the focus is now on 3D PET where the size of the image is 11–15 times larger than in 2D (after exploiting symmetries). Thus, there is continuing need for new ideas in image reconstruction algorithm development. Finally, it is worth mentioning that the explosion in the use and utility of the Internet provide some resources of specific interest to the medical imaging community.

Bibliography

Amari, Shun-ichi. Back propagation and stochastic gradient descent method. *Neuro-computing*, 5(4–5):185–196, 1993.

Darken, Christian and John Moody. Towards faster stochastic gradient search. In NIPs, volume 91, pp. 1009–1016, 1991.

Deshpande, H. S., K. J. Karande and A. O. Mulani, Area optimized implementation of AES algorithm on FPGA, *2015 International Conference on Communications and Signal Processing (ICCSP)*, 2015, pp. 0010–0014, doi: 10.1109/ICCSP.2015.7322746.

Gupta, P. and S. L. Tripathi, Low power design of bulk driven operational transconductance amplifier, 2nd International Conference on Devices for Integrated Circuit (DevIC), no. 4, pp. 241–246, 2017.

Jacobs, Robert A. Increased rates of convergence through learning rate adaptation. *Neural Networks*, 1(4):295–307, 1988.

Jadhav, M. M., G. H. Chavan and A. O. Mulani, Machine learning based autonomous fire combat turret, *Turkish Journal of Computer and Mathematics Education* (TURCOMAT), 12(2), 2372–2381, 2021.

Jin, Wen, Zhao Jia Li, Luo Si Wei, and Han Zhen. The improvements of bp neural network learning algorithm. In WCC 2000-ICSP 2000. 2000 5th International Conference on Signal Processing Proceedings. 16th World Computer Congress 2000, volume 3, pp. 1647–1649. IEEE, 2000.

Kulkarni, Priyanka and A. O. Mulani, Robust invisible digital image watermarking using discrete wavelet transform, *International Journal of Engineering Research & Technology* (IJERT), 4(1),139–141, 2015.

Liu, Zhengjun, Aixia Liu, Changyao Wang and Zheng Niu. Evolving neural network using real coded genetic algorithm (ga) for multispectral image classification. *Future Generation Computer Systems*, 20(7):1119–1129, 2004.

Madan, Akansha and Divya Gupta. Speech feature extraction and classification: A comparative review. *International Journal of Computer Applications*, 90(9), 2014. doi: 10.5120/15603-4392

Mendiratta, N. and S. L. Tripathi, 18nm n-channel and p-channel Dopingless asymmetrical junctionless DG-MOSFET: Low-power CMOS based digital and memory applications. *Silicon* 14:6435–6446, (2021). https://doi.org/10.1007/s12 633-021-01417-5

Mulani, A. O. and P. B. Mane, An efficient implementation of DWT for image compression on reconfigurable platform, *International Journal of Control Theory and Applications*, 10(15):1–7, 2017.

Mulani, A. O. and P. B. Mane, High throughput area efficient FPGA implementation of AES algorithm, *Intech Open Access Book on Computer and Network Security*, February 2019.

Randall Wilson, D. and Tony R. Martinez. The need for small learning rates on large problems. In IJCNN'01. International Joint Conference on Neural Networks. Proceedings (Cat. No. 01CH37222), volume 1, pp. 115–119. IEEE, 2001.

Shinde, Ganesh and Altaf Mulani, A robust digital image watermarking using DWT- PCA, *International Journal of Innovations in Engineering Research and Technology* (IJIERT), 6(4):1–7, 2019.

Swami, S. S. and A. O. Mulani, An efficient FPGA implementation of discrete wavelet transform for image compression, International Conference on Energy, Communication, Data Analytics and Soft Computing (ICECDS), 2017.

Wen, Shiping, Shuixin Xiao, Yin Yang, Zheng Yan, Zhigang Zeng and Tingwen Huang. Adjusting learning rate of memristor-based multilayer neural networks via fuzzy method. *IEEE Transactions on Computer-Aided Design of Integrated Circuits and Systems*, 38(6):1084–1094, 2018.

Zeiler, Matthew D. Adadelta: an adaptive learning rate method. arXiv preprint arXiv:1212.5701, 2012.

Chapter 14

Leveraging health care industry through medical IoT

Its implementation and case studies within India

Mohamed AbuBasim, S. Kaliappan,
V. Shanmugasundaram, and S. Muthukumar

CONTENTS

DOI: 10.1201/9781003335801-14

14.1 INTRODUCTION

The health care industry is an amalgamation of various sectors within the eco-system that offers products and services to people who need therapeutic, rehabili-tative, precautionary, or palliative care. It encompasses the production and sale of items and services that contribute to the maintenance and restoration of health among patients. It is segmented into several sectors and sub-sectors according to services like telemedicine, products like medical devices and surgical instruments and finance like medical insurance [1]. The sector is highly dependent on multiple disciplines ranging from trained professionals who work towards the pursuit of good health for individuals and the population, which contributes to more than 10 percent of GDP to the overall developing country's economy. Different aspects contribute to the process of establishing and implementing reform in health care. But convoluted and inefficient government regulations are undoubtedly playing a major role in addition to ecological and scientific variables. Furthermore, disease patterns, physician demographic characteristics, and technological breakthroughs all contribute to changes in our entire health care system. As our society changes, our health care needs also evolve around it. The current scenario is projected to be characterised by rising market demand, increased pricing, and more consumer awareness. These reforms will result in a positive shift in the health care industry's landscape. The 1990s had a slow pace of growth, and health expenditures per capita touched an all-time low. The ponderous nature of growth might be characterised by the efficiency with which numerous health initiatives were executed [2].

Digital transformation and availability of mobile internet have been the major trends in recent years, particularly in the health care domain. During the COVID-19 scenario, they played a major role in virtually assisting every patient's need, to book vaccination slots in nearby hospitals or primary health care centres based on the daily availability, to provide oxygen concentrators, ventilators to the needy, monitoring oxygen levels, and other vital parameters for the patients at home. Both e-health and m-health systems are governing the whole health care eco-sphere, which includes online consultation, delivery of medicines, diagnostic technologies, and remote monitoring. This pandemic made us view the health care sector in a different light in all the aspects. The epicentre of the sector is to enable it by various cutting-edge disruptive innovations that are provided by pioneering health care startups.

Additionally, certain startup health tech portals help in maintaining digital health care records of the patients by creating a longevous database of patient's history [3]. This helps in eliminating the errors made by hospital administration personnel and provides early alerts to avert serious medical emergencies. The significance of IoT devices and their influence on the health care sector cannot be overemphasised as they have pushed forward many tech giants like Apple, Google, Samsung, and many others to invest in bringing out a new generation of digital health applications to monitor blood-glucose levels, EEG, ECG, heart rates, and sleeping patterns. Many health care organisations like WHO,

Centers for Disease Control and Prevention, American Medical Association, Indian Medical Association, and Food and Drug Administration have started to adopt and explore several avenues to improve the quality of treatment and disease diagnostic techniques for the people worldwide. These solutions contribute to the transformation of the public health care paradigm by automating routine procedures and resulting in better treatments. The mission of enabling technology in the health care sector is to create a channel for community development and critical human services.

14.1.1 Challenges and opportunities

Technology has taken over many sectors through digital transformation by helping them to achieve a tremendous amount of success through the years. But, when it comes to the healthcare industry, still it is reluctant to utilise the full advantage of its opportunities. A report shows that only 7 percent of health care organisations have transitioned to digital, although the transformation is 15 percent in other sectors. The digitalisation of the medical landscape has historically been slow. Until COVID-19, the health care sector lagged in its use of digital techniques. COVID-19 catalysed hospitals, pharmaceutical businesses, and diagnostic firms to digitally access $200 billion of India's $400 billion health care budget and revealed the important role of technology [4]. The health care industry's indisputable growth is mainly due to the deployment of connected devices in the fields of remote health care monitoring and diagnostic technologies. India's health care sector had gained wider coverage in global media during to the second wave of COVID-19 (especially between April and June 2021), which affected several lakhs of people. There is a considerable deficit in the health care ecosystem between the available physicians and the needs of people who get affected. Inadequate medical equipment, medications, COVID-19 care centres, hospitals, life-saving pharmaceuticals, and immunisations have all contributed to India's current plight. As a growing nation with the world's second largest population, India's health care system requires significant investment to radically transform its health care delivery environment and pave the path for cutting-edge medical technologies. Nowadays, hospitals concentrate more on the use of digital platforms to improve patient care delivery, remote monitoring, and care continuum management need assistance. COVID-19 (especially the second wave) has exposed inadequacies in primary care with physician consultations serving as the sole conduits. As a result, virtual consultations, telemedicine, remote patient monitoring, home health care, e-pharmacy, e-diagnostics, and tele-ICU services are all needed to expand to all the villages and cities in the future.

Security breach of patient records is a major threat to IoT-based systems. All these devices are obliged to collect retain and communicate their prescriptions, consultation records, and personal data, which are highly sensitive. If this technology is embedded in all health care sectors, government policymakers and regulatory bodies must ask the developers and manufacturers of the medical devices to continuous monitor their data analytics policy. This is the only way; the patients can trust and embrace the use of devices in their daily life. The priority of the developers is to ensure the patients' anonymity and their privacy [5]. Similarly, they are responsible for ensuring that the equipment is shielded with a stringent firewall that the hackers couldn't have access to that equipment from any part of the world. More significantly, these IoT-based devices should be used based on individual's needs and must be maintained properly. Patients must be aware of why and what type of their data is being used for further analysis. The professionals must perpetuate integrity and prevent misuse of data by others. The manufacturers must provide enough training to their experts who use IoT devices to guarantee that the "user behaviour is not to blame for data exploitation". This is a far more realistic possibility than a global health care strike. There are many hurdles during the evolution of IoT technology, but it has revolutionised the health care industry in endless, comprehensive, and compelling phenomenal breakthroughs for both professionals and patients.

14.1.2 Present and future investments in digital health

India spends the least amount of public expenditure on health care among the BRICS (Brazil, Russia, India, China, South Africa) nations. A more complex approach to finance adaptation, which is described in Figure 14.1 in the health care system, is needed. Accessing technology is considered a major issue for people in rural areas. It is highly limited in the fields of telemedicine and diagnostic technologies. The technology growth lies in developing them, which can be utilised by the population at the bottom-most pyramid than those at the apex. As stated by renowned economist Jeffrey D. Sachs, this is the moment for massive investment in the health care sector like China did for four decades by allocating large-scale budgets [6]. The COVID-19 crisis had shown the concept of the corporatist idea of government failure in endeavouring to maximise the people's well-being. Instead of direct public expenditure, policies have targeted restructuring all economic sectors and stimulating the economy via the banking markets and monetary advantages. The COVID experience demonstrates that the current fiscal policy approach is in desperate need of improvement. Privatisation is highly motivated only by profit maximisation and capital formation. Even during this pandemic, corporate earnings have grown to a larger extent as compared with the rest of the economy. The main problem that persists in the Indian health care system is "privatisation", which leads to bias in offering health benefits and access to allocated resources [7].

Developing and improvising the infrastructure means providing the surplus amount of oxygen supply through exclusive machinery and concentrators, setting up ventilator units, and not about spending a large expenditure on glorifying the hospital rooms with beautiful canvas and air-conditioning them. The government sector is withdrawing its efforts with the growth of the private sector. The government opted to expend on medical insurance instead of expenditures on enhanced health care facilities for the impoverished and encouraging physicians to work in rural and urban areas in India. The central point is that the private health care system will focus on the poor until the collective insurance scheme persists. Grants provided towards the needy and weaker sections of society is one of the strategies aimed at ensuring equality. However, data suggests that without robust evaluation and monitoring procedures, it is inequitable to improvise the mistakes committed by the government against the health care system. Under the Atma-Nirbhar Bharat Abhiyan scheme, the Indian government has launched a special monetary and exhaustive package for the health care sector, which covers 10 percent of India's GDP [8]. This is completely focused on enhancing the development of public health and its reforms to confront the future challenges.

Due to new variants in lifestyle, there is a higher possibility of emerging infectious and contagious diseases, higher health care costs leading to demand for medical devices, development in telemedicine, e-health, and m-health benefits. The government's initiatives with tax relaxations made India invest in the health care sector to a major extent when compared to others during this pandemic scenario. Some of the major statistics are gleaned from the "Invest India Portal" [9]. In 2020, the Indian health tech business was estimated to be worth $1.9 billion. By 2023, it is predicted to reach $5 billion, representing a compounded annual growth rate (CAGR) of 39 percent. With this investment, the Indian health care sector is predicted to be among the top three in the global market by 2022. Information technology in the health care sector is predicted to double in size from its present value of $1 billion by 2022. By 2022, the health care diagnostics industry is predicted to increase at a CAGR of 20.4 percent to achieve $32 billion, which was previously $5 billion in 2012. The widespread need for online platforms that provide consultations to patients who are situated in distant locations, such as rural or distant areas, is further intensified in the face of shortages of medical experts [10]. By 2022, the Indian telemedicine industry is predicted to register a CAGR of 20 percent to $32 million, which was previously $15 million. It is expected that, by 2030, the National Digital Health Blueprint has the potential to generate over $200 billion in additional economic value for the health industry.

In India, where the health care sector accounts for 80 percent of the entire market, there is a strong investment preference from both global and local investors. The

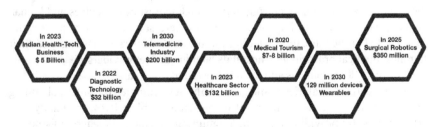

Figure 14.1 Recent predictions of Indian health sector and their major investments.

hospital sector revenues are projected to rise from $61.8 billion in 2017 to $132 billion by 2023, with a CAGR of 16–17 percent. According to industry experts, the Indian medical tourism market is anticipated to expand from $3 billion to $7–8 billion by 2022. A billion dollars' worth of diagnostics facilities is presently available in India. The overall diagnostic sector consists of almost 25 percent of companies that are small and mid-sized (15 percent in labs and 10 percent in radiology). The present valuation of the primary care sector is at $13 billion. This figure does not include the organised sector, which makes up a very small percentage of the total industry. As many as 70,000 "Ayushman Bharat Centres", which are designed to provide basic health care services to communities located within a closer distance of their homes, are operating in India. Health insurance is the second biggest part of the non-life insurance sector, and it adds up to 20 percent of non-life insurance revenues. Gross direct premium revenue, covered by health insurance increased 17.16 percent a year through $6.87 billion in FY 20. Increasingly, India is developing as a very significant market for the wearables, with roughly two million devices sold in 2017, and that number is predicted to reach almost 129 million devices by 2030. AI is used to track the patient's overall health and to provide recommendations on the best possible therapy at the proper moment. According to industry analysts, the surgical robotics market of India is expected to record a CAGR of 20 percent (2017–2025) to a total of $350 million by 2025.

14.2 FUNCTIONAL FRAMEWORK OF THE MEDICAL IOT OR HEALTH CARE IOT

The Industry 4.0 era has made it feasible to remotely monitor and control all of the IoT-enabled devices in the health care industry, unlocking a tremendous amount of potential to ensure patient safety and wellness while allowing doctors to provide cutting-edge treatment. Enrolment in Medicaid has also boosted the level of patient participation and satisfaction as visits to physicians have become more reliable and simple. Consequently, remote monitoring of a patient's health helps to reduce the time of stay in the hospital, as well as the chance of being readmitted. The number of devices being connected to the Internet of Things will also help in lowering health care expenses and improving treatment results. Internet of Things (IoT) plays a constructive role in promoting improvements in the health care business by considering the characteristics of devices and human contact, which is key to implementing health care solutions. There are several uses for IoT in health care, including the benefits it offers to patients and their families, medical practitioners, hospitals, and insurance providers.

The recent developments and the widespread use of wearables to monitor blood pressure, heart rate, calories burned, and daily steps completed result in obtaining tailored data for the users. It has also impacted people's lives, especially elderly patients, by providing real-time monitoring of vital parameters. Disruption in their daily routine triggers an alert to their caregivers, which is considered as one of the major milestones in the elderly health care sector. IoT empowers health care providers to be more conscientious and connects them with patients in a proactive manner. The data obtained from IoT devices is beneficial to doctors in determining the type of treatments based on various recommendations that are most appropriate for their patients and concerning predicted results. The medical equipment, which is connected using IoT, is tagged with sensors that help in the real-time tracking of essentials like defibrillators, oxygen concentrators, and nebulisers. The power of IoT helps in deploying the physicians at various sites within the hospitals or primary and secondary health centres (during the current pandemic scenario) to treat the patients. Data gathered through health monitoring devices like a fitness band and wrist watches can be used by insurance companies for further process and claims [11]. By compiling and analysing this data, the insurance providers can maintain some transparency to identify fraud allegations and discover candidates for endorsing purposes and also in the pricing and risk management process. The fact that insurers may provide incentives to their clients for utilising and contributing health data collected by IoT devices encourages consumers to buy them and participate in IoT-based health research.

14.2.1 Technologies involved in building the framework of medical IoT

The mode of IoT framework used for health care applications combines the technologies of IoT technology and cloud computing. It includes standard protocols to collect some of the patients' information from sensors and sophisticated equipment to a designated health care system [12]. The configuration of a medical IoT is the grouping of multiple elements of an IoT network that are in a health care setting and organised in a way that is coherent with the health care ecosystem. The system model comprises three main elements: a publisher, a broker and a subscriber [13]. To increase the effectiveness of this network of interconnected sensors and other devices, the publisher refers to a collection of medical devices that might separately or concurrently collect the patient's important information. This data could include vital parameters such as blood pressure, heart rate, temperature, oxygen saturation, electrocardiogram, electromyogram and many more. This is followed by transmitting the information collected from in publisher to the broker. The broker is responsible for processing the information and uploading it to the cloud for further analysis. Lastly, the subscriber is responsible for visualising the above information using smart devices like mobile phones, watches, laptops or tablets [14].

14.2.1.1 Identification technologies and their standards

As long as the above conditions are met, the publisher can evaluate this data and provide feedback on any signs of anomalies in physiological status or health deterioration in the patient. This is done by joining discrete elements together into one hybrid grid, in which each component is devoted to serving a particular function mostly on IoT network and cloud [15]. Although there is nothing to propose a common structure for this technology, since the architecture for an HIoT relies on health care needs and application, the generic structure is not necessarily well-defined. Substantially, several changes have been made in medical IoT configuration to achieve the objectives of physicians and other researchers in this field. The medical IoT network is a designated one for specific diseases, so flexibility is an important criterion in designing the configuration without compromising its ethical guidelines in diagnostic procedures.

Due to the patient's data being available from the designated node (sensor), which might be located at remote locations, developing a health care IoT system should be taken into account for its accessibility. With the proper identification of the nodes and sensors in the health care network, it is possible to implement an effective pinpointing of the nodes and sensors that are available in the health system. The identification procedure consists of issuing a unique identifier (UID) per authorised entity so that it can be readily recognised and unambiguous data interchange can be obtained. When everything in the health care system is organised around a digital ID, every resource inside the system (such as hospitals, doctors, nurses and carers) comes with a unique digital ID. This assures the authentication of the assets and the connectivity among them in a digital environment. For information, specific tools for identification have indeed been interpreted within the context [16]. The Open Software Foundation (OSF) has devised two types of identifiers: a universally unique identifier (UUID) and a globally developed unique identifier (GDUID). UUID, a component of the distributed computing environment (DCE), may be used without requiring hierarchical supervision [17]. In a health centre system, the sensors and actuators are assigned and acknowledged individually, which helps to ensure the effective working of the system. However, there is a possibility that the distinctive component identification of an IoT system may evolve over the entire lifecycle of the IoT system owing to the ongoing upgrades of IoT-based solutions. For the device to ensure the coherence of the health care system, the device must have had a method to keep up to date with its relevant information. Because modifying the configuration may influence several changes in the network components along with the faulty diagnosis. Besides that, the adoption of IoT in health care facilitates the utilisation of new technologies capable of (1) retrieving data using a global identification number, (2) safely overseeing the identity of the components utilising distinct encryption and authentication techniques and (3) constructing a universal database search for identifying the IoT services with the UUID scheme.

14.2.1.2 Communication technologies and their protocols

Most communication methods allow entities in a health care IoT network to be in frequent touch with one another. The technologies can be generally categorised as "short-range and medium-range communication" technologies. The short-range communication technologies, such as Bluetooth and near-field communication (NFC), are protocols being used to establish a link among the assets in a limited range or a body area network (BAN), whereas the medium-range technological advances, such as Wi-Fi and cellular data, support communication for an even more distance, such as communication between a base station and the root/parent/central node of a BAN. Short-range communication ranges from few centimetres to several metres, and long-range communication is measured in millimetres and metres. Some of the most extensively used communication methods are RFID, Wi-Fi, Zigbee, Bluetooth and satellites,

RFID (radiofrequency identification) encompasses a tag and a reader, which is used for a shorter range of communication of 10 cm to 200 m. The tag is made from a microchip and an antenna to represent a particular item or device in the IoT ecosystem. The reader interacts with the tag using radio waves, which is susceptible to transmitting information from the item. The use of RFID allows health care practitioners to find and monitor health equipment in an instant. One of the key advantages of RFID is it doesn't need any external energy source. Nevertheless, it is a rather vulnerable protocol and might potentially have incompatibility concerns when synchronised with a smartphone. In addition to its main usage in mobile phone accessories, Bluetooth technology implements a short-distance wireless communication protocol that implements UHF (ultrahigh frequency) radio waves. Two or more medical equipment can be interfaced using this technology, which has a frequency range of 2.4 GHz and a range of 100 m. It possesses lower interference, as a result of low energy consumption and the transmitted data is highly encrypted for authentication purposes, but it fails in health care applications in a long-range communication.

Zigbee is one among other mainstream communication protocols used to link medical equipment, transfer information and relay results between them. This possesses characteristics similar to Bluetooth technology with a frequency range of 2.4 GHz, but a better communication range than Bluetooth. The significance of the Zigbee protocol is low power consumption, higher transmission rate and network capacity due to its "mesh network topology". Near-field communication (NFC) protocol is a type of technology that works under the principle of electromagnetic induction between two loop antennas that are positioned so close together. Whenever NFC devices are in an active state, they may perform several operations. The passive mode radiofrequency transmitter uses just one device, which transmits the radiofrequency whereas the second device works as a receiver. Both devices may create radiofrequency energy concurrently in the event of active mode, and it is possible to communicate data without coupling [18]. Wireless Fidelity (Wi-Fi) is an example of a wired local area network (LAN) that implements the IEEE 802.3 standard. In contrast to Bluetooth, which only has a transmission range of between 30 and 40 feet, the range delivered by Wi-Fi is at least 70 feet. It is widely used in all hospitals due to its good compatibility with all electronic devices, security and control features. The major drawback of using a Wi-Fi network is high power usage and its performance fails inconsistency due to its network issues. Satellite communication has proven to be remarkably successful and beneficial in faraway and widely isolated geographic areas like villages, mountains and seas where other communication protocol fails. Satellites receive the signals from the ground and then amplifies them, before sending the signals back to earth. It is exclusively used for its highlighted characteristics like fast data transmission, rapid broadband access, long-term consistency and a high degree of compatibility. Unfortunately, the amount of power required for satellite transmission is far more than that required for other communication modalities.

14.2.1.3 Location technology

The location technology helps in identifying and tracks the location of all the assets inside the health care network, allowing the appropriate staff to reach them. The most significant part is that this technology helps in maintaining the record of the treatment process based on the availability of the resources. The global positioning system (GPS) is a widely used technology used for tracking, but not limited to navigation and surveying. The use of satellites helps to keep track of the assets. For an asset to be identified through GPS, there must be an unobstructed clear view between the asset and four separate satellites. In health care IoT, GPS may be used to find the location of ambulances, health care practitioners, patients, and so on. Furthermore, GPS can be used only in outside environments because the infrastructures of the environmental landscape may serve as a barrier to transmit the signal between the asset/object and the satellite. Local positioning system (LPS) is being used when there is a "localisation principle" to follow a moving item by using the radio signal, which is transmitted by the asset/object while in motion to a network of pre-deployed receivers [19]. LPS can be implemented by using other short-distance communication technologies like Bluetooth, Zigbee and RFID, but ultra-wideband (UWB) is used due to its higher temporal resolution, which allows the receiver to properly calculate the time of arrival. Some researchers have used UWB-based localisation systems, which can use time difference of arrival (TDOA) [20] [21], relative and degree of difference time for arrival [22] and so on for tracking. When GPS is conjoined with other higher bandwidth communication technologies, it can be explored for several avenues of health care technologies.

14.3 APPLICATIONS OF IOT IN THE HEALTH CARE SECTOR

The health care sector provides a larger scope for adopting newer and breathtaking technologies using a Web-based interconnected universe, which is defined as the "Internet of Things". Through IoT, patients are provided with the option of staying with doctors through remote access and virtually visiting them based on their needs. The technology also helps in tracking the assets in hospitals and allocating resources during the time of emergency. This also helps in facilitating the patients with chronic illness during surgery and diagnostic analysis, tracking their medical reports for examination. By automating patient workflow, it minimises inefficiency and errors by reducing the cost. Besides medical items and the Internet of Things, health care is also using the technology in different ways, extending from telemetry to sensor integration to medical device connections. An increase in the number of connected devices for the IoT in health care may be seen in the application of medical equipment, such as ECG and EEG monitoring, delivery of medicine during the surgery and monitoring of blood glucose level. There are numerous applications designed to serve as a solution for both the patient and the health care provider.

14.3.1 Remote monitoring of the patients by medical practitioners

While health care practitioners are currently well prepared to collect data from patients while they are in the hospital, patients must eventually be discharged and travel home with little or no degree of clinical care. If individual health devices can create data that is readily accessible to the hospital's clinical system at that point, they may serve as a remote communication network for a patient who has been released. The data collected by these IoT devices and how they get processed is explained in Figure 14.2. This may therefore serve as an early warning system for health care practitioners, assisting them in avoiding an expensive and catastrophic crisis in readmissions to the hospital's facility, if vital signs deteriorate [23]. The primary scenario is the provision of long-term remote patient monitoring for chronic patients. The majority of patients with chronic conditions

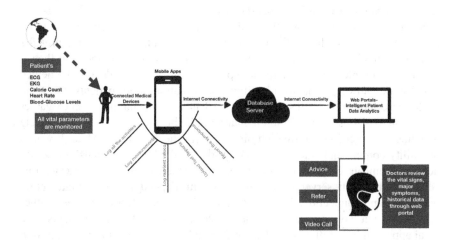

Figure 14.2 Basic architecture of remote monitoring systems in healthcare using IoT framework.

are between the ages of 70 and 85 and may be observed remotely by health care specialists that specialise in remote health monitoring. The primary advantage is easy accessibility: it decreases expenses, minimises travel for older and delicate patients and saves the physician's time. Additionally, long-term remote patient monitoring with this system can result in improved interventions; so that if something disconcerting is found in the data, the patient can be called to the clinic to see a specialist. If nothing is detected, there is no need for unnecessary routine visits that can be conducted remotely [24].

The technology aims at providing 'zero touches' configuration, which means that patients do not need to install any app to dynamically send their medical history to their health care provider; they do not want to enter any of their authentication details into an app; and it's not needed to pair a medical device [25]. All unnecessary, time-consuming device configuration is completely removed, making it very simple for health care providers to install such devices as well as for patients to employ them at home. The ability to monitor patients from patient's home is a major benefit for a lot of folks. Monitoring the patient remotely enhances the patient comfort by allowing susceptible groups to remain at home – for instance, even during the COVID-19 pandemic, remote monitoring might have allowed hospitals to gather a large amount of data from individuals at home without compromising their health and protecting them from virus transmission. The key factor here is that the patient's information is highly secured. We harnessed the compliance standards and their capabilities of current clinical systems: if such systems were not secure and adhered to fundamental security and privacy standards, the firms that depend on them would come to an end, since protecting patient information is their primary value in their business. The patient's information is never disclosed with a device vendor; it's never been stored in any server (cloud) except the medical database in which the device is associated with a patient. This is referred to as integrity and confidentiality [26].

Some devices depend on Bluetooth, in which suppliers attach a Bluetooth module to a patient's house through a smartphone application: this requires the patient to supply their contact information when installing the app, and then it becomes a complex procedure to encrypt their data [27]. The data in the device cloud is devoid of personally identifying information about patients. Only the health care industry remains dependent on Bluetooth communication to gather data at the patient's home. Within the automobile business, for example, vehicle manufacturers get IoT information directly from the vehicle's SIM card; within the utility commercial enterprise, water grids installed in individuals' houses do not communicate over Bluetooth; instead, the smart metre communicates directly with the utility server. Both of these systems depend on constant interaction with the cloud connection. The health care industry may learn from these other businesses that have previously implemented large-scale IoT deployments. The ideal approach is to follow their lead and include direct-to-cloud IoT connection into the device. In this regard, medical IoT is exceedingly secure and far safer and perhaps less susceptible to cyber-attacks than Bluetooth-enabled equipment.

It has becomeclear that remote health care techniques will be the predominant method of choice mostly in the future.

There are many stages of remote care: the first, which was accelerated substantially by the outbreak of the pandemic, is tele-consultation, in which the doctor and patient connect through a phone or video. The second level requires working with all of the telemedicine firms that grew exponentially during the crisis to include remote monitoring devices that continuously collect patient-generated health data. We could expect rapid growth in telemedicine technology as a promising launching pad for future emergencies: monitoring the patient remotely facilitated by the massive implementation of medical IoT devices, permitting health care authorities and health care providers to be ready for the next pandemic, which will once again confine large numbers of people to their homes.

14.3.2 IoT in vaccine manufacturing and supply chain management

Indeed, no other challenge the globe is facing currently is more critical than the swift manufacturing and deployment of the various vaccinations available to battle COVID-19. The faster vaccination doses can be manufactured, distributed and its supply chain management system is clearly sketched in Figure 14.3. The sooner the world can return to a peaceful transition and people may live fearlessly once again. As has been the case with most manufacturing in recent years, the Internet of Things is proving to be a critical element in the process when it comes to developing vaccines. Customarily, the timeframes to have a vaccine out from the door are 10–40 percent for quality management and 50–70 percent for the immunisation and biotech sector. Given the critical nature of time sensitivity in the context of the COVID-19 vaccination, the potential to shorten these lead periods would enable the manufacture of larger dosages [28]. This would result in the future saving of further lives.

Consequently, companies are moving towards IoT, automation, ML and DL algorithms to ramp up production efficiency. The usage of IoT, AI and machine learning could also significantly increase production plant efficiency, especially during a time when it becomes unsafe to have the majority of staff in one facility owing to social distance rules. IoT sensors enhance the reliability and regulation of critical process control parameters, such as temperatures, pH values, pressure, oxygen levels, and so on [29]. With information from these categories, AI may select and analyse the relevant data to forecast a performance degradation before it occurs. Additionally, IoT sensors may be utilised to ensure the safety of on-site employees by tracking ranges and air circulation. The information obtained through IoT devices that are analysed by AI will help in contributing to the sustainable improvement in processes, even while the operation is underway. These remarkable improvements will enable those manufacturing COVID-19 vaccines to drastically increase the production, resulting in the vaccination of even more people in much lesser time.

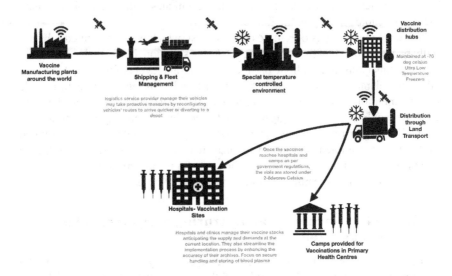

Figure 14.3 Supply chain management involved in vaccine manufacturing and how IoT is used in delivering the vaccines from manufacturing plants to hospitals and other vaccination sites.

The Internet of Things is becoming more valuable to the vaccination campaign in several ways, not only in factories. Implementation, transportation and preservation are all provided with further hurdles, as several commercially approved vaccines lose efficacy if not kept chilled. Temperatures monitoring is critical since the Pfizer vaccine should be stored at −70°C and Moderna's vaccine must be at −20°C. However, since IoT sensors have already been used to monitor performance measures such as temperature, assuring that perhaps a vaccine does not deteriorate due to heat, which is a trivial issue of configuring the specifications to meet the needs of the vaccine being kept or carried. Temperatures may be remotely controlled through IoT, with alerts sent if any problems happen or performance drops below the standard. In these instances, the logistics service provider managing their vehicles may take proactive measures by reconfiguring vehicles' routes to arrive quicker or diverting to a depot where experts may be prepped and ready to perform essential repairs before vaccinations are contaminated. IoT benefits the COVID-19 routine immunisation in more ways than just one since IoT devices could have been used to assist the vaccinated people [30]. Wearable devices may be used to monitor when people receive their first shot and to remind them when it becomes due for their second. Similarly, hospitals may use IoT to streamline the implementation process by enhancing the accuracy of their archives. Lastly, IoT sensors can be used in temperature monitoring systems, which involves the secure handling and storing of blood plasma. Plasma contains antibodies that are used to develop novel vaccines and more efficient therapies.

14.3.3 Evolution of IoT in wearable technology

Smart wearable technology has permeated the system to the point that smart watches are considered newsworthy, and the growth of wearables is moving at a rapid rate. Wearable technology usage will have more than quadrupled in the previous years, driven by a market desire for self-monitoring. According to Business Insider Intelligence analysis, more than 80 percent of the customers are open to wearables [31]. This increased appetite for wearables has created a booming economy, and insurance companies are finally realising the purpose of having wearables for their consumers and employees.

Wearables in the health care industry involve the use of digital devices worn by consumers, such as Fitbits and smartwatches that are used to track health information like calorie consumption, sleeping habits, pulse rate and walking distance of the users, which can be seen in Figure 14.4. Smart wearables, like Apple Watch and Fitbit, can monitor a user's heart rate, which is beneficial during their exercise activity. However, implications in cardiology research are still in progress. The Apple Heart Study, which established the efficacy of smartwatch-based atrial fibrillation screening, is one of the most possibly the best studies on wearable health care technology to date. They are now collaborating with Johnson & Johnson on a large-scale research study to not only screening for but also diagnose and manage the illness. The Heart Rhythm Society's professional cardiologists have ought to embrace the technology. Wearables may benefit in the early detection and

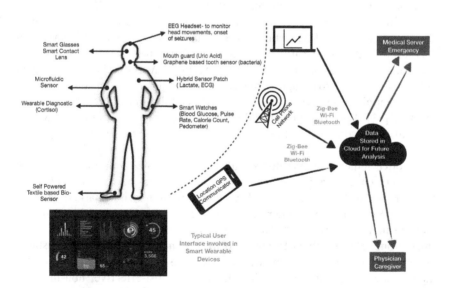

Figure 14.4 Major wearable sensors used for monitoring the vitals and how it is tracked by the physician. The figure also depicts the simple user interface of smart wearable devices.

treatment of illnesses, although continuous screening is highly discussed among the health care industry [32].

Nowadays, sleeping patterns are often tracked using mobile applications and smartwatches in which wearables are widely used. We can analyse our sleep habits and target specific points of contention for progress with this technology. Materialistic fitness devices are being commercially marketed to monitor moderate and REM sleep cycles. For instance, "Withings ScanWatch" is perhaps the first Food and Drug Administration (FDA)-approved wearable device capable of detecting sleep apnoea and other anomalies. There lies a thin line of barrier between standard diagnostic forms/questionnaire and wearables in detecting major vital parameters. Telemedicine is becoming increasingly popular in big hospitals, aided by an estimated increase in the number of FDA-approved devices. Although this technology could help us to gain better insights into a sleep schedule, patterns and habits, it is improbable that it will replace standard practices of sleep testing shortly. Many patients and families with hereditary chronic disorders such as diabetes are anxious about their medication. Medical wearables address this issue in a multitude of situations. They may assist with tracking blood sugar levels, reminding patients to take insulin, and giving appropriate nutrition and lifestyle counselling. A classic insulin pump is a valuable tool since it is intended to manually release medicine to manage their blood sugar levels based on the desired level. Recently, engineers have developed comprehensive "closed loop systems," which constantly detect and modify a patient's blood sugar levels in real time. Closed-loop insulin administration systems integrate continuous glucose monitoring (CGM) sensor with an insulin infusion pump equipped with a computerised algorithm for infusion rate adjustment. Meanwhile, Klue, a startup, has developed behavioural surveilling software that utilises AI to identify the patient's food habits and provide a deeper perspective into their behavioural patterns. Implementing an autonomous system to recognise the daily food habit and insulin administration may potentially transform a simplified and healthier diabetic routine in their lifestyle [33]. With the advent of IoT, asthmatics can now pinpoint warning signs of an episode before the start of the attack. Asthma frequently exhibits itself as abrupt exacerbations that may occur anywhere at any time. Patients may notice impending symptoms with the ADAMM asthma monitor before their occurrence. This is a splotchy wearable that may be worn on either the front or rear of the upper torso. The gadget monitors vital markers such as heart rate, respiration rate and respiratory rate, looking for anomalies that may indicate an impending attack. This information is then transmitted to the user's smartphone, alerting them to the need to respond. Additionally, this asthma monitor has inhaler detection, which enables users to identify and monitor their inhaler usage. Investment in efficient and cost-effective, non-invasive and easy to access ubiquitous wearable technology should be emphasised for entrepreneurs and corporations in this sector.

14.3.4 IoT in surgery: a digital revolution

Operation theatre is a place where a patient is surrounded by several doctors. It's jarring to gaze over there at dozens or more monitors specialised to solitary medical equipment without considering how the Internet of Things (IoT) may considerably ease the constraints of coordinating so many systems. It's easy to see how the medical business, and hospitals, in particular, will account for a sizeable portion of Cisco's $19 trillion Internet of Things market potential by 2022. The digital revolution in surgery has already occurred, and it is groundbreaking as a result of a novel blend of IoT, big data sophisticated analytics insights and intelligent medical equipment. A sample framework involved in surgical IoT is described in Figure 14.5. Thousands of individuals endure cardiac arrhythmias as a result of cardiac disease. Such irregularities manifest as a frantic heartbeat that is very disruptive and may result in potentially dangerous seizures and heart attacks. While a few medicines can alleviate symptoms, they do little to address the underlying condition known as atrial fibrillation. "CardioThings," a pseudo name of the startup, is tackling the issue via ablation, which involves gently burning off lesions with a laser.

This procedure involves introducing a catheter further into the atrium to attempt ablation of the AFib-causing lesions. Each component is hardwired to a monitor, which displays a view of the interior of the heart through video streaming from the catheter's tip [35]. However, unlike many devices, the data does not cease

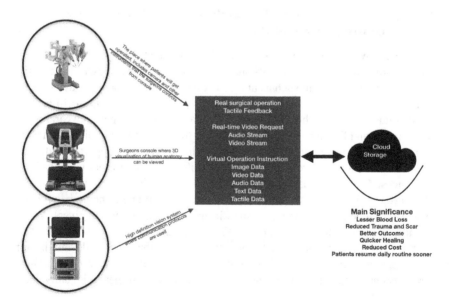

Figure 14.5 Framework involved in typical IoT-based surgery system. As a reference, Da-Vinci Surgical Robotic System [34] is used.

between both the heart and the screens. CardioThings collaborates with two real-world Internet of Things stalwarts, PTC ThingWorx and Glassbeam, to enable something even more powerful. ThingWorx simulates the catheter's function to transfer encrypted data to the cloud, where Glassbeam [36] can examine it. Glassbeam converts fragmented data into a format data in the form of understandable insights that the equipment maker may utilise to help surgeons perform better during a surgical intervention. CardioThings and other manufacturers of high-value assets may also benefit from this kind of data by increasing the dependability of their catheter device. Others may utilise IoT Insights to improve the reliability of CAT scans and MRIs by detecting when even the tiniest component is exhibiting indications of failure and facilitating a repair that keeps the equipment operational. Consider CardioThings' optical catheter, which is tiny enough to pass safely via a vein, accessing the heart and imaging it to identify the abnormalities causing AFib. The surgeon may then use CaridoThings' monitors to define the bounds of the lesions to evaluate which are withering and therefore should be burnt away. The lesions are subsequently removed by the laser beam, which is attached to the sensor-embedded catheter, and the patient is healed. The maintenance time for million-dollar equipment like MRI and CT are not only pricey for the hospitals since it is not charging patients, but it also hinders patients from receiving the best treatment possible. ThingWorx allows other devices to connect via the cloud and once their fragmented data is in place, Glassbeam's analytical engine may integrate and recombine it to find any anomalies.

14.3.5 IoT technology for Alzheimer's, Parkinson and dementia patients

For persons with Alzheimer's as well as dementia, technology has grown more user-friendly. While elevated tools such as GPS trackers can observe and assist them, minimal equipment such as unique eating utensils or apparel may also certainly make living with dementia easier. This can be seen in Figure 14.6 for a briefer understanding of how IoT technology can be used as a part of wearable for Alzheimer's and Parkinson patients. Likewise, there are "virtual voice assistants" such as the Amazon Echo or Google Home that really can support a multitude of situations, ranging from prescription notifications to virtually controlling appliances. The major onset symptoms of Alzheimer's in the elderly is roaming. Their progressive cognitive impairment renders them prone to being bewildered [37]. This vulnerability imposes a significant burden on their caregivers. When an individual is treated with Parkinson's disease, they begin to lose control of their subcortical structures. Such patients possess imbalance while moving or standing. Correlation between their brain and other organs get deteriorated. Thus, caregivers are anticipated to stay alert and attentive to assist them with daily practice in this instance.

IoT technology embedded in their daily wearables such as wristbands, socks, coats, buttons, spectacles and shoe bottoms may be quite beneficial through RFID

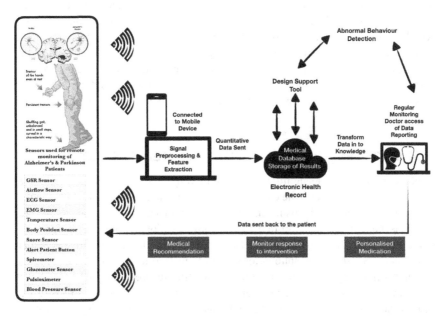

Figure 14.6 IoT technology involved in monitoring Alzheimer's and Parkinson patients.

chips, GPS trackers and motion sensors [38]. When a patient leaves their home or during their routine work, their IoT sensors alert their caregivers or nurses. This way, you can monitor your loved ones' locations while they are suffering memory issues. One of the most common complications of Parkinson's disease is immobility in their movements. Their physical parts cannot move whenever the mind is prepared to do a certain action. In this case, the patient harms themselves as a result of a disruption in the central nervous system. Now, Internet of Things sensors can monitor a patient's daily regular movements as well as other physiological movements and behaviours [39]. The system monitors the patient's routines for just about any odd or abrupt changes and alerts the caretaker to the potential hazard.

During these circumstances, IoT-enabled solutions enter the equation. Technological advancements alleviate these issues by giving doctors specific data that empowers them to make knowledgeable real-time judgements. Consequently, clinicians may devote more time to sensitive elements of the patient's health and concentrate on the critical ones. The above three cases in the health care fields where IoT may have a significant influence. Clinicians may get concise and exhaustive information about a medical illness through the evaluations of many other doctors. Through critical meta-data analytics, health care workers may have a better understanding of patient rehabilitation. For instance, IBM Watson has built a discovery tool called "Knowledge Integration Toolkit (KnIT)". It is a three-step process that includes the study, investigation and elucidation of health care processes. KnIT reads and analyses scientific publications using Watson's AI platform. Watson frees physicians and health care professionals from data processing, enabling them to focus on improving patients' conditions and vital signs [40]. Biotricity [41], for example, is a pioneer in the development of remote patient monitoring technologies. This firm develops medical gadgets that bring together real-time patient's information and firmly send it to the cloud for further analysis.

14.4 SUCCESSFUL CASE STUDIES OF MEDICAL IOT

Health care has become significantly more intelligent as a result of technological advancements. IoT pushes the envelope of inpatient care to several breakthroughs. It streamlines the mechanisms involved in health care professionals. Several hospitals have indeed committed to focusing on global investments for technology adoption. It's difficult to redefine a sector with deeper IoT prospects than health care. Sensor-based innovations provide numerous advantages on a range of perspectives, spanning from giving treatments to preventing chronic illness and diagnostic techniques. The intelligent system with IoT-based software can acquire and evaluate data from patients autonomously, pinpointing the illness-influencing factors and assisting in preventing it at the earliest. Medical IoT may assist in enhancing diagnostic accuracy by continuous monitoring and controlling the vital parameters. In health care, IoT capabilities help the patients to interact with physicians during this pandemic scenario. Some successful case studies and implications of medical IoT is explained in the following section.

14.4.1 Apple Watch Series' ECG feature saved human life

The electrocardiogram (ECG) function of the Apple Watch saved the life of a 61-year-old man in March 2020. R. Rajhans, a retired pharmaceutical executive, felt sick and proceeded to examine his ECG through Apple Watch Series 5 provided to him by his son. Siddharth, his son, stated that his father started experiencing arrhythmia signals, or abnormal heartbeats, through the night. After the outcomes were examined, it was revealed that his father had a weak ejection fraction and needed immediate cardiac surgery to fix his mitral valve. The procedure was delayed somewhat because of the pandemic, during which Rajhan's ECG was monitored through his Apple Watch. The purpose of an Apple Watch would be to achieve the right balance between technology use and health consciousness in line with the priorities of people's life. Numerous lives have been saved worldwide as a result of the Apple Watch's functionalities such as ECG, step count, calories consumed, sleeping patterns and more. The watch, in conjunction with the ECG app and the irregular heart rhythm alert capability, may assist users in diagnosing AFib symptoms that result in a stroke [42].

14.4.2 Telemedicine platforms to access health care for all people

Prayagraj, Uttar Pradesh, based startup Sprint Medical concentrates more on telemedicine and patient care. Sprint Medical offers economical medical attention – anytime, anywhere – with elevated digital technologies. Sprint Medical empowers patients to search for and schedule arrangements with the finest physicians from across India. They provide augmented doctor consultations under which a nurse assists patients throughout the approach and use connected gadgets that facilitate physicians to do virtual diagnostics in a more precise manner. Sprint Medical collects the real-time vital parameters like ECG, EKG and blood glucose level, which can be tracked, examined and monitored and their medical history can be saved for future analysis. IoT has made remarkable growth in various domains of health care like consultation, radiology, pathology and pharmacies. It is predicted to propel the Indian telemedicine industry to $5.5 billion by 2025 [43]. As a result, many telemedicine platforms are emerging in our country like Practo, Lybrate and Phable, It is a basic right for all individual to access primary health care. The future is here, and it has unified patients and physicians. All data is stored on the electronic database, which can be accessed through several IoT-enabled platforms with different frameworks, which are seen in Figure 14.7, that are critical to the development of this health care ecosphere in India.

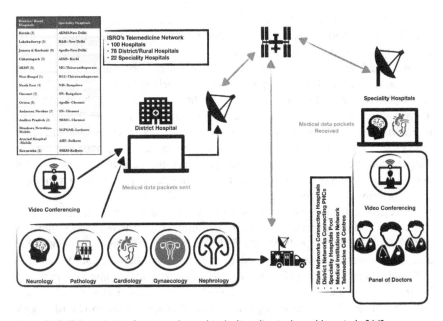

Figure 14.7 Telemedicine framework used in Indian district/ rural hospitals [44].

The Indian Space Research Organisation (ISRO) pioneered telemedicine in India in 2001 with a Telemedicine Pilot Project through a collaboration with Chennai's Apollo Hospitals by connecting the Apollo Rural Hospital in Aragonda village in Andhra Pradesh's Chittoor district. Government agencies like ISRO, the Department of Information Technology (DIT), the Ministry of External Affairs, the Ministry of Health and Family Welfare and various state governments collectively contributed important strides in establishing telemedicine services in India [44]. Few more worthwhile instances of the efficaciously documented telemedicine facilities in India encompass mammography services at Sri Ganga Ram Hospital, Delhi and oncology at Regional Cancer Centre, Thiruvananthapuram [45]. Telemedicine is extremely effective in sites where a tremendous crowd of people converge every once in a while and availability of public care becomes a necessity; for illustration, the Government of Uttar Pradesh uses telemedicine during Maha Kumbhamelas [46]. Telemedicine is indeed a discipline that has been effective in piquing the corporate sector's interest and compelling it to participate actively in public health programmes. Narayana Hrudayalaya, Apollo Telemedicine Enterprises, Asia Heart Foundation, Escorts Heart Institute, Amrita Institute of Medical Sciences and Aravind Eye Care are considered as major players in the telemedicine sector with a major collaboration with state governments and ISRO, the organisation which contributes with updated technology [47].

14.4.3 IoT in robotic-assisted surgery

Dr. Naresh Trehan of Delhi's Escorts Heart Institute and Research Centre (EHIRC) employed a robotic arm coupled with an endoscopic camera to execute a sophisticated coronary bypass surgery in 2002 [48]. The arm is comprised of three black chopstick-like arms interconnected by cables and equipped with pivots and pistons [49]. Dr. Apurva Vyas in Sterling Hospitals, Ahmedabad, accomplished the first robotic surgery in India and the third in the world for a 37-year-old patient with median arcuate ligament syndrome with the assistance of a four-armed Da-Vinci robotic system. Through this, it was possible to reach the celiac artery's root, whereas human fingers never could [50]. The three-dimensional elevated picture enlarged the artery multiple times, assisting the doctor in achieving precision and minimising unintended destruction to healthy tissue, arteries and nerves.

Dr. Tejas Patel, a renowned cardiac surgeon based in Gujarat, executed robotic surgery for a mid-aged patient with a ruptured aneurysm while seated 32 kilometres distant from the client, probably the first time surgery has been conducted from a distant location. The whole procedure may also have been accomplished with a 20 mbps internet speed [51]. This will redefine not only coronary intervention, but the whole vascular system. It has the potential to influence the livelihoods of hundreds of people in rural areas. With yet another spectacular model of robotic surgery, specialists at Chennai's SIMS Hospital used a robotic arm to undertake knee restoration surgery. Dr. Vijay Bose, the orthopaedic surgeon who directed the procedure, commented of its success, that a single arm helps in the entire surgery [52]. The system notifies the surgeon when there is a little change in the angle of twist. Amrita Institute of Medical Sciences in Kochi recently reached a milestone by conducting 500 robotic surgeries. The Gynaecological Oncology Department began using robotics in 2015 and has already accomplished numerous gynaeco-logical cancer surgeries and other procedures. From the early part, people under-going robotic surgery asserted benefits such as rapid recovery, less pain, the fewer amount of blood loss and very lesser complications [53]. Furthermore, the cost of surgery was kept as least expensive at AIMS, so that everyone can get benefit in India.

14.5 CONCLUSION

India's tremendous digital disruption has enabled the Indian population to gain access to all government services. The vision of establishing India as a truly digital democratic nation is making major headway, thanks to the improvised online infrastructure and internet connectivity. As the government intensifies its deployment of data analytics, IoT and cloud technology, innumerable spheres such as education, health care, crime prevention, law enforcement and public safetywill witness more development. Digital breakthroughs in technologies are contributing more in redefining patients' data, system optimisation, the elimin-ation of adverse error, the betterment in clinical care and the reduction of health expenditures. By deploying virtual care using telemedicine, smart wearables and uninterrupted network connectivity, it would be easy to provide superior treatments to the entire population. But from the other perspective, the private enterprises have reaped the economic benefits of IoT and big data computing investments. Organisations have gleaned meaningful actionable intelligence using cloud data platforms, minimising the costs and increased product reliability by generating sustainable business models and offering a great customer experi-ence. Since numerous health care devices are networked through IoT, tracking and observation will help in improving patient care, and the information collected will facilitate the delivery of democratic health care statistics through which basic requirements of the entire population can be accommodated. This helps in maintaining the health of the entire population, resulting in a booming economy. The battle has unquestionably commenced, and a fragment of this evolution is now accessible by everyone. The Indian government has taken a massive collab-orative push with all startups and defence organisations to advance to the next stage of the revolutionary path towards digital health care. The financial sector of our country should provide a big picture for all the enterprises by investing in infrastructure, technical expertise and strengthening the hospital–industry cooper-ation to stimulate a breakthrough of technological developments that enrich the whole country. Together, we can lay the groundwork for a much more economic-ally vibrant, diverse and enlightened India.

References

[1] Anthony Ledesma, C. M. et al. (2014). *Health Care Sector- Overview.* Retrieved May 28, 2021, from Washington State University website: https://s3.wp.wsu.edu/uploads/sites/606/2015/02/SectorOverview_HC_Spring2014.pdf

[2] Watch, E. (2010, June 29). *Health Care Industry.* Retrieved May 28, 2021, from *Economy Watch* website: www.economywatch.com/world-industries/health-care

[3] Bhardwaj, T. (2021, May 26). Technology, digital health solutions to address overall healthcare ecosystem: Dr Rajesh Gupta, MyHealthcare. *Financial Express.*

[4] Dhingra, B. (2021, May 28). Is healthcare industry still reluctant to take full advantage of technology? *BioSpectrum.*

[5] IoT.Business.News. (2021, May 25). *IoT and the Healthcare Revolution.* Retrieved May 29, 2021, from *IoT Business News*: https://iotbusinessnews.com/2021/05/25/64714-iot-and-the-healthcare-revolution/

[6] PTI. (2021, May 13). India's public spending on healthcare lowest in BRICS nations; tech can improve access: DEA Secy. *The Economic Times.*

[7] Chakravarty, A. (2021, May 29). Time for maximum welfare. *The Tribune.*

[8] Sharma, N. C. (2021, January 27). India to roll out sustained investment in healthcare delivery system: Vardhan. Retrieved May 30, 2021, from livemint website: www.livemint.com/news/india/india-to-roll-out-sustained-investment-in-healthcare-delivery-system-vardhan-11611767674291.html

[9] Bajaj, A. (2020). Working towards building a healthier India. Retrieved May 29, 2021, from Invest India – National Investment Promotion and Facilitation Agency: www.investindia.gov.in/sector/healthcare

[10] Bora, N. (2020, October 23). The future of healthcare investment in India. *Outlook.* www.outlookindia.com/website/story/opinion-the-future-of-healthcare-investment-in-india/362801

[11] Rajashekhar Karjagi and Manish Jindal. (2020). What can IoT do for healthcare? Retrieved May 30, 2021, from Wipro Limited: www.wipro.com/business-process/what-can-iot-do-for-healthcare-/

[12] Oryema, B. (2017). Design and implementation of an interoperable. *14th IEEE Annual Consumer Communications & Networking Conference (CCNC)* (pp. 45–52). Las Vegas, NV: IEEE.

[13] Ahad, M. T.-L. (2019). 5G-based smart healthcare network: architecture, taxonomy, challenges and future research directions. *IEEE Access, 7,* 100747–100762.

[14] Hanji, M. N. (2020). Internet of things based distributed healthcare systems: a review. *Journal of Data, Information and Management, 2.*

[15] Kadhim, K. T. (2020). An overview of patient's health status monitoring system based on internet of things (IoT). *Wireless Personal Communications, 114,* 1–28.

[16] Scholtz, J.-Y. L. (2002). Ranging in a dense multipath environment using an UWB radio link. *IEEE Journal on Selected Areas in Communications, 20,* 1677–1683.

[17] H. Aftab, K. G. (2020). Analysis of identifiers in IoT platforms. *Digital Communications and Networks, 6* (3), 333–340.

[18] G. Cerruela Garc'ıa, I. L.-N. (2016). State of the art, trends and future of Bluetooth low energy near field communication and visible light communication in the development of smart cities. *Sensors,* 1968.

[19] Sichitiu, R. P. (September 2006). Angle of arrival localization for wireless sensor networks. *3rd Annual IEEE Communications Society on Sensor and Ad Hoc Communications and Networks* (pp. 374–382). Reston, VA: IEEE.

[20] Young, D. P. (2003). Ultra-wideband (UWB) transmitter location using time difference of arrival (TDOA) techniques. *The Thirty-Seventh Asilomar Conference on Signals, Systems and Computers*, (pp. 1225–1229). Pacific Grove, CA.

[21] Zetik, R. (2004). UWB localization-active and passive approach (Ultra-Wide BAnd Radar). *21st IEEE Instrumentation and Measurement Technology Conference (IEEE Cat. No. 04CH37510)* (pp. 1005–1009). Como, Italy: IEEE.

[22] Gunderson, R. J. (2002). Ultra-wideband precision asset location system. *IEEE Conference on Ultra Wideband Systems and Technologies, (IEEE Cat. No. 02EX580)* (pp. 147–150). Baltimore, MD: IEEE.

[23] Rai, N. (2020, November 20). *Reimagining India's digital transformation in vital sectors*. Retrieved June 5, 2021, from livemint website: www.livemint.com/opinion/ online-views/opinion-reimagining-india-s-digital-transformation-in-vital-sectors-11606199247553.html

[24] ETHealthWorld (2021, June 3). Digital Healthcare, let us adapt and transform. *Economic Times – Healthworld supplement.*

[25] Ramaswamy, R. (2017, July 22). Riding technology: The role of IoT in healthcare surveillance. *The Economic Times – Rise supplement.* https://economictimes.indiati mes.com/small-biz/security-tech/technology/riding-technology-the-role-of-iot-in-healthcare-surveillance/articleshow/59710658.cms?from=mdr

[26] Ahmad, S. (2021, May 11). Startups lean on remote-use devices to boost Covid-19 patient care at home. *Business Standard.* www.business-standard.com/article/ companies/startups-lean-on-remote-use-devices-to-boost-covid-19-patient-care-at-home-121051100871_1.html

[27] Scott, J. (2021, June 2). How remote patient monitoring can provide insightful patient care. Retrieved June 5, 2021, from Health Tech Website: https://healthtechmagazine. net/article/2021/06/ata2021-how-remote-patient-monitoring-can-provide-insightful-patient-care

[28] Mahihenni, H. (2020, May 20). How IoT and AI could help speedup and scaleup Covid-19 vaccine manufacturing process? Retrieved June 2, 2021, from LinkedIn website: www.linkedin.com/pulse/how-iot-ai-could-help-speedup-scaleup-covid-19-hadi-mahihenni

[29] Mopidevi, R. (2021, February 5). IoT-enabled process validation system for COVID-19 vaccine rollout. Retrieved June 2, 2021, from *Control Engineering*: //www.con troleng.com/articles/iot-enabled-process-validation-system-for-covid-19-vaccine-rollout/

[30] Sisto, A. (2021, February 3). The role of IoT in scaling up vaccine manufacturing. Retrieved June 2, 2021, from tern PLC website: www.ternplc.com/blog/the-role-of-iot-in-scaling-up-vaccine-manufacturing

[31] Phaneuf, A. (2019, July 19). Latest trends in medical monitoring devices and wearable health technology. Retrieved June 2, 2021, from *Business Insider-India*: www. businessinsider.in/latest-trends-in-medical-monitoring-devices-and-wearable-hea lth-technology/articleshow/70295772.cms

[32] IoT Business News. (2020, April 6). the future of IoT in healthcare: wearable technology. Retrieved June 2, 2021, from *IoT Business News* website: https://iotbusinessn ews.com/2020/04/06/90470-the-future-of-iot-in-healthcare-wearable-technology/

[33] Katiyar, R. (2021, May 10). How wearable tech insights are improving healthcare? Retrieved June 2, 2021, from TechGenyz website: www.techgenyz.com/2021/05/10/ how-wearable-tech-insights-are-improving-health-care/

[34] Intuitive (2019, 3). About da Vinci Systems – Surgical robotics for minimally invasive surgery. Retrieved June 7, 2021, from Intuitive website: www.davincisurgery.com/da-vinci-systems/about-da-vinci-systems

[35] Nordlinger, C. (2015, May 5). The Internet of Things and the operating room of the future. Retrieved June 2, 2021, from Medium website: https://chrisnordlinger.medium.com/the-internet-of-things-and-the-operating-room-of-the-future-8999a143d7b1

[36] Pandit, P. (2020, January 2). Glassbeam technology – innovation at its peak for medical device manufacturers. Retrieved June 2, 2021, from Glassbeam website: www.glassbeam.com/blogs/glassbeam-technology-innovation-at-its-peak-for-medical-device-manufacturers/

[37] Jackson, R. (2020, April 30). IoT and healthcare technologies converge for better patient care. Retrieved June 4, 2021, from readwrite website: https://readwrite.com/2020/04/30/iot-and-healthcare-technologies-converge-for-better-patient-care/

[38] Dementia Care Central. (2020, June 29). Guide to assistive technology, aids and apps for persons with dementia and caregivers. Retrieved June 4, 2021, from Dementia Care Central website: www.dementiacarecentral.com/caregiverinfo/technology/

[39] Verse Technology (2017, November 2). Using IoT to help patients with Alzheimer's and Dementia. Retrieved June 4, 2021, from Medium website: https://medium.com/@VERSETechnology/using-iot-to-help-patients-with-alzheimers-and-dementia-2c643dff312b

[40] IBM Newsroom (2014, August 28). IBM Watson ushers in a new era of data-driven discoveries. Retrieved June 4, 2021, from IBM Newsroom webiste: https://newsroom.ibm.com/2014-08-27-IBM-Watson-Ushers-in-a-New-Era-of-Data-Driven-Discoveries

[41] Biotricity Newsroom. (2017, September 13). Remote patient monitoring: A telemedicine solution to chronic disease. Retrieved June 4, 2021, from Biotricity website: www.biotricity.com/remote-patient-monitoring-telemedicine-solution-chronic-disease/

[42] Tech Desk (2020, October 21). Apple Watch Series 5's ECG feature saves life of a 61-year-old Indian man. *The Indian Express*.

[43] Brand Desk (2021, May 10). Telemedicine is changing healthcare and saving lives in India. Retrieved June 3, 2021, from DNA India – Health website: www.dnaindia.com/health/report-telemedicine-is-changing-healthcare-and-saving-lives-in-india-2889671

[44] ISRO (2005). *Telemedicine – Healing Touch Through Space*. Bangalore: Publications and Public Relations Unit, ISRO Headquarters.

[45] Sudhamony, K. N. (2008). Telemedicine and tele-health services for cancer-care delivery in India. *IET Communications*, 2 (2), 231–236.

[46] Mishra S.K., (2004). Telemedicine application in maha kumbhmela (Indian festival) with large congregation. *Telemed J E Health*, 107–108.

[47] Dasgupta A, D. S. (2008). Telemedicine: A new horizon in public health in India. *Indian J Community Med.*, *33*, 3–8.

[48] Waldman, A. (2003, May 18). Indian heart surgeon took talents home. *The New York Times*.

[49] Analytics India Magazine (2019, April 5). How surgical robot assistants are becoming A reality in Indian hospitals and healthcare sector. *Analytics India Magazine*.

[50] Business Standard staff (2017, April 24). Ahmedabad surgeon performs India's 1st robotic surgery. *Business Standard.*

[51] *Economic Times* (2018, December 6). Meet Dr. Tejas Patel – The Indian cardiologist who repairs woman's heart from 32 km away. *Times Now News.*

[52] PTI (2020, April 2). Tamil Nadu's first robotic knee replacement surgery performed at SIMS Hospital. *The Week.*

[53] Express News Service (2019, March 23). Amrita Institute of Medical Sciences creates history in robotic surgery. *The New Indian Express.*

Chapter 15

Energy-efficient architectures for IoT applications

Smrity Dwivedi

CONTENTS

15.1 INTRODUCTION

After blooming of a new era in computation, Internet of Things (IoT) has emerged as a basic building block of all computing. IoT is a smart technology that interconnects each and every "thing" through a network in one form or to another form. The term "thing" includes components like sensors, actuators, hardware, software, and storage spread over multiple disciplines such as health care, industry, transport, home appliances, and science and technology. The main objective of IoT is to maximize the communication of hardware objects with the physical world and to convert the data that are harvested by these objects into useful information without using any human aid. IoT consists of three elements: hardware, middleware, and presentation. The hardware element is comprised of battery-powered embedded sensors, actuators, and communication systems. The sensors collect data from the monitoring area, and their communication hardware sends the

DOI: 10.1201/9781003335801-15

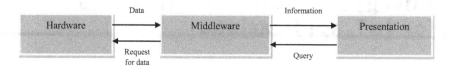

Figure 15.1 Elements of IoT.

collected data to the middleware element. An enormous amount of data received by middleware is processed and analyzed by using various data analysis tools to extract interpretable information. The presentation element of IoT is responsible for the visualization of processed data and results in a novel and easily readable form. It also receives user queries and passes them to the middleware element for necessary actions as in given Figure 15.1. The figure shows the elements and data transfer in IoT systems. The limited battery power of hardware elements is consumed while collecting and transmitting data to other blocks. The more are the data collected and analyzed, the more the accuracy of the extracted information but, at the same time, the more is the energy consumed. Due to energy limitations, there is a need to maintain a trade-off between quality of information extracted and energy consumption by IoT systems. Hence, the lifetime of any resource in IoT depends upon the energy present. The loss of energy affects the whole environment under observation. Thus, there is, on a priority basis, need to reduce energy consumption for the prolonged lifetime of resources and the effective operation of IoT systems. So, a hierarchical architecture is proposed in this chapter to improve the energy efficiency of IoT. The proposed architecture exploits the fact that IoT resources consume negligible energy in the sleep mode. Henceforth, the architectural design allows the sensors to switch to sleep mode under three scenarios: first, when it is not necessary to sense the target environment in a given period of time; second, when the coverage area can be compromised for battery life; and third, when the battery level is critically low. Apart from it, when the sensors are in sleep mode, the proposed architecture allows the allocated middleware resources to either switch to sleep mode or get released and reprovisioned later for energy efficiency. Thus, the resources of hardware and middleware elements of IoT have been "tuned" together in the proposed architecture for better performance and energy saving. To achieve energy saving, the main objectives of the proposed architecture can be stated as follows: (1) to provide a mechanism for the efficient energy utilization of both hardware and middleware elements of IoT systems; (2) to predict and control the sleep interval of sensors depending upon their previous usage history and remaining battery level; and (3) to reprovision the allocated cloud resources when corresponding sensors are in sleep mode.

Energy is as valuable resource for IoT network because the devices used for IoT applications are low-power devices (battery operated). In some applications, the nodes are placed in remote areas to support hardware. When battery of the node drains out its power, it is difficult to replace the battery.

The first way to save energy is to place the nodes in an efficient way. The network that provides an effective communication between energy-constrained devices can be considered as a suitable network for IoT. The energy efficiency of the network is directly related to network lifetime. By consuming the energy in balanced way, the network lifetime can be prolonged. In the IoT network, the traffic patterns are one to one or many to one and one to many. In this type of traffic, many to one (nodes to base station) traffic pattern happens most of the time. So the nodes near the sink will be overloaded. They carry their own data as well as they forward the data of other nodes, which leads to huge energy depletion and quick node death. This problem is referred to as the energy hole problem. Once this energy hole occurs, the entire network will get disconnected and this severely affects the lifetime of entire network. In most of the literature, it is stated that huge amount of energy is not utilized due to quick network disconnects (energy hole). In this chapter, a hierarchical placement of relay nodes are proposed by considering the data traffic and a suitable routing mechanism is implemented to avoid non-uniform energy drainage. In the given architecture, relay nodes are placed in a hierarchical way. Relay node is similar to a sensor node. It performs all other tasks (communication, computation, etc.) except sensing. In the present work, sensors are responsible for communication, such as route computation and data transmission from sensor nodes to relay nodes. The reason for introducing the relay node is to reduce the data burden (overload) and computational complexity of sensors, which prevents sensors from quick battery drain out. Relays are placed one hop neighbor to the sensor node to improve the connectivity of the network. In the proposed network architecture, both efficient placement of nodes and routing mechanism are combined to improve the performance of the network. The placement of the relay node is done based on the data traffic of the network. In a high data-traffic area, one relay node is assigned for one sensor. In a medium data-traffic area, one relay node is assigned for two sensors and in a low-data traffic area, one relay node is assigned for three sensors. Also relay node is responsible for routing; it does optimal path selection from the source to destination (sensors to sink). The relay node finds the optimal path by considering the residual energy of the node. Because for wireless sensor network, resource constraints are handled by low-power devices, which are naturally constrained. When a neighbor node is bad in energy level, the path will be disconnected and data loss may occur, hence residual energy is considered as an important metric to compute the energy-efficient path. The combination of hierarchical placement of nodes and energy-balanced routing increases the network lifetime.

15.2 ROLE OF SENSOR NODES AND RELAY NODES

In the presented network architecture (Figure 15.2), sensor nodes are responsible for sensing, computation (processing) and communication (transmission and reception). Relay nodes are responsible for computation and communication. Sensor senses (collect the environmental data), processes the sensed data, and transmits the data to one hop relay node whereas the relay node collects the data

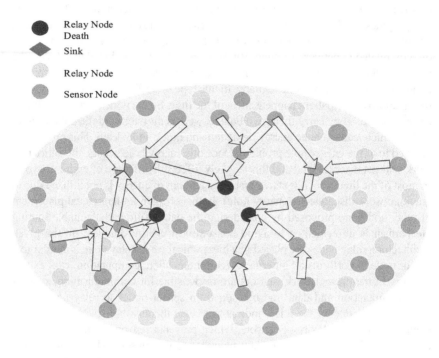

Figure 15.2 Random relay placement.

from sensors and other relays. After data aggregation, the relay node transmits the data to the sink. Flooding process is handled by the relay nodes. Relay floods the control packet (route request or route replay) and finds the energy-efficient path to the sink. That's why sensing process and routing process are split to reduce the complexity of the node. The sensor will be free from path computation process and relays will be free from the sensing process. The energy level of the relays is kept high when compared to sensors, because communication complexity is more in relays, when compared to sensors (Table 15.1).

15.3 HIERARCHICAL NODE PLACEMENT

After considering the energy hole problem, efficient relay placement is done in the given network architecture. The assumption is that the radius of the network is 50 meters. The relay nodes are far from the sink. Consider they are 40 meters away from the sink. They will carry only the sensor data, the burden of relay in this area will be less. The relay nodes, which are placed at a moderate distance to the sink, considering 20 meters, will carry one hop sensor data and as well forwarded data of relays placed at 40 meters. Hence the burden of relay will be medium in this area. The relay nodes, which are near the sink, just consider 5 meters, will carry

Table 15.1 Role of sensors and relays

Role of the node	Sensor	Relays
Sensing	Yes	No
Path computation	No	Yes
Data processing	Yes	Yes
Transmitting	Yes	Yes
Receiving	Yes	Yes
Communication to sink	No	Yes

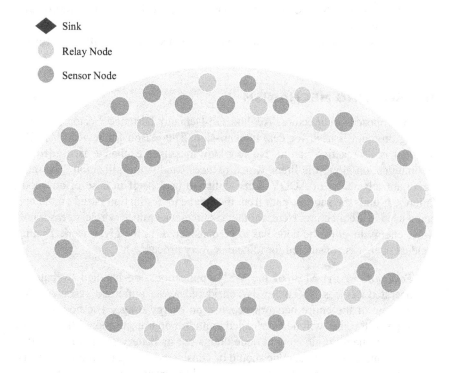

Figure 15.3 Hierarchical relay placement.

one hop sensor data and as well forward data of relay nodes, which are placed at 40 meters and 20 meters. This leads to heavy node burden and quick energy depletion. In Figure 15.2, the red color indicates death nodes due to overburden. To solve this problem, the density of relay nodes are increased toward the sink, and the ratio of sensor and relays are varied with respect to traffic area in proposed network architecture (hierarchical). Relay nodes are kept one hop neighbor to sensors and relays. Sensors are placed in a random fashion based on application requirements. The relay node carries the data from one hop sensor neighbor and one hop relay neighbor. In Figure 15.3, the red circle indicates high data traffic area, blue circle indicates medium data traffic area, and green circle indicates low data traffic area. By considering the traffic area, the relay nodes are assigned to sensor nodes. The ratio of sensor and relay are described based on traffic area.

The second important reason for varying the relay node density is implementation cost because relay nodes have a high cost (high battery capacity). Hence the proposed architecture satisfies both data traffic and network cost.

The following basic node placement assumptions are made.

1. For every sensor node, one relay node is assigned in a high-traffic area (red circle).
2. For two sensor nodes, one relay node is assigned in a medium-traffic area (blue circle).
3. For three sensor nodes, one relay node is assigned in a low-traffic area (green circle).

15.4 ROUTING MECHANISM

In the proposed architecture, Ad hoc On-Demand Distance Vector (AODV) routing protocol is used for data transmission. The reason for selecting AODV protocol is its reactive nature. No topology messages exchange is needed for communication along the links, which reduces bandwidth utilization. The most important advantage of AODV is its ability to heal itself in case of any node failures. It finds the shortest path from the source to destination based on the hop count used in architecture. Here, for the resource-constrained wireless sensor network, energy level of the node has to be considered. In the given work, routing residual energy is considered for route discovery process.

(a) *Residual energy*: Most of the WSN applications are handled by battery-operated devices, so energy is considered as an important resource. The lifetime of the entire network depends on energy usage. The nodes which are near the sink will be overloaded in multi-hop transmission. This leads to uneven energy drainage and node drains out its battery soon. To avoid this problem, energy of the node should be considered during the route discovery process. The nodes with good energy level can be considered as intermediate nodes from the source to destination.

(b) *RREQ packet format*: AODV protocol uses route request (RREQ) packet for route discovery from the source node to destination node. To implement the R in AODV, it should be added to the RREQ control packet.

(c) *Route selection by the destination node based on R value*: The route selection of AODV protocol is done by the destination node. Whenever the destination node receives the route request, it discards further route request and starts sending the route replay to the source. Figure 15.4 (flow chart) presents the route selection procedure of destination node.

$$RE = E_r/E_{max} \qquad (15.1)$$

Type	Flags	Reserved	Hop Count
RFEQ (broadcast ID)			
Destination Address			
Destination sequence number			
Original address			
Original sequence number			
Residual Number			

Figure 15.4 RFEQ packet format.

Figure 15.5 explains the route selection of the destination node based on residual energy. It selects the node, which has good residual energy. After starting the RREP timer, the destination node sends reply RREP to each RREQ packet stored in the cache. After data transmission, it removes all the entries in the cache.

(d) *Basic assumptions*: Stationary nodes (relays and sensors) are placed. Sensors are placed randomly and relay nodes are placed in a hierarchical fashion. Nodes are aware of the residual energy information. The battery levels of relay nodes are high when compared to sensor nodes. Relay nodes are placed one hop neighbor to sensor node and relay node. Sink is not limited by energy.

15.5 ENERGY STORAGE THROUGH IOT ENERGY-EFFICIENT ARCHITECTURE

The integration of renewable energy and optimization of energy use are key enablers of sustainable energy transitions for mitigating climate change. Modern technologies such the Internet of Things (IoT) offer a wide number of applications in the energy sector, that is, in energy supply, transmission, and distribution, as well as demand. IoT can be employed for improving energy efficiency, increasing the share of renewable energy and reducing environmental impacts of the energy use. This chapter includes the existing literature on the application of IoT in in energy systems and in the context of smart grids particularly. Furthermore, enabling technologies of IoT, including cloud computing and different platforms for data analysis, are discussed. Furthermore, the challenges of deploying IoT in the energy sector are reviewed, including privacy and security. Some solutions to these challenges such as blockchain technology is given. This survey provides energy policymakers, energy economists, and managers with an overview of the role of IoT in optimization of energy systems.

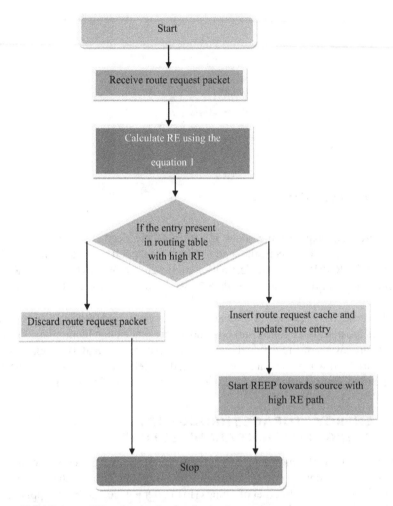

Figure 15.5 Minimum RE path selection by destination node.

15.5.1 Concepts

Industrial revolutions have been divided into four phases. In the first revolution, new sources of energy were discovered to run the machines and the mass extraction of coal and the invention of steam power plants were significant development stages in this phase. The second revolution known as mass production and electricity generation, which was a period of rapid development in industry, distinguished by large-scale iron and steel production and during this phase, many large-scale factories with their assembly lines were established and formed new businesses. The third revolution introduced computers and the first generation of communication technologies, such as telephony system, which enabled automation in supply chains. The fourth industrial revolution included a wide variety of modern technologies such as communication systems (e.g., 5G), intelligent robots, and the Internet of Things (IoT). IoT interconnects a number of devices, people, data, and processes by allowing them to communicate with each other amazingly. Hence, IoT can help improve different processes to be more quantifiable and measurable by collecting and processing large amount of data. IoT can potentially enhance the quality of life in different areas like medical services, smart cities, construction industry, agriculture, water management, and the energy sector. This is enabled by providing an increased automated decision making in real-time and facilitating tools for optimizing such decisions.

The global energy demand rose by 2.3 percent in 2018 compared to 2017, which is the highest increase since 2010 and CO_2 emissions from the energy sector hit a new record in 2018. Compared to the pre-industrial temperature level, global warming is approaching 1.5°C, most likely before the middle of the twenty-first century. If this trend prevails, the global warming will exceed the 2°C target, which will surely have a severe impact on the planet and human life. Environmental concerns such as global warming and local air pollution, scarcity of water resources for thermal power generation, and the limitation of depleting fossil energy resources raise an urgent need for more efficient use of energy and the use of renewable energy sources. Many studies have shown that a nonfossil energy system is almost impossible without efficient use of energy and/or reduction of energy demand, and a high-level integration of residual energy sources, both at a country level, regional, or globally. On the basis of the United Nations Sustainable Development Goals agenda, energy efficiency is one of the key drivers of sustainable development. Also, energy efficiency offers economic benefits in long term by reducing the cost of fuel imports/supply, energy generation, and reducing emissions from the energy sector. For increasing energy efficiency and a more optimal energy management, an effective analysis of the real-time data in the energy supply chain plays a key role. The energy supply chain, from resource extraction to delivering it in a useful form to the end users, includes three major parts: (i) energy supply including upstream refinery processes; (ii) energy transformation processes including transmission and distribution of energy carriers; and (iii) energy demand side, which includes the use of energy in buildings, transportation sector, and the industry. Figure 15.6 shows these three parts with their relevant components. IoT employs sensors and communication technologies for sensing and transmitting real-time data, which enables fast computations and also optimal decision-making. Also, IoT helps the energy sector to transform from a centralized to a distributed, smart, and integrated energy system. This is a key requirement in deploying local, distributed residual energy sources, such as wind and solar energy, as well as turning many small-scale end users of energy into prosumers by aggregating their generation and optimizing their demand whenever useful for the grid. IoT-based systems can automate, integrate, and control processes through sensors and communication technologies. Large data collection

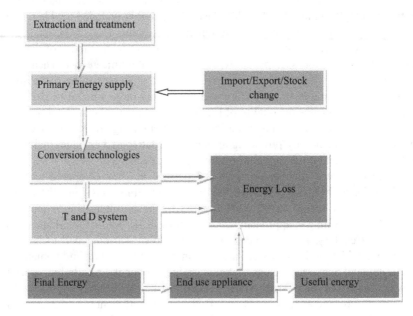

Figure 15.6 Energy supply chain.

as well as use of intelligent algorithms for real-time data analysis can help to monitor energy consumption patterns of different users and devices in different time scales and control that consumption more efficiently.

While planning an IoT application, which is the first step in designing IoT systems, the selection of components of IoT such as sensor device, communication protocol, data storage, and computation need to be appropriate for the intended application. Like, an IoT platform planned to control heating, cooling, and air conditioning (HVAC) in a building requires utilizing relevant environmental sensors and using a suitable communication technology as shown in Figure 15.7 for the different components of an IoT platform. IoT devices, which are the second components of the IoT platforms, can be in the form of sensors, actuators, IoT gateways, or any device that joins the cycle of data collection, transmission, and processing as an IoT gateway device enables routing the data into the IoT system and establishing bidirectional communications between the device-to-gateway and gateway-to-cloud.

The communication protocols that are the third component of the IoT platform enable the different devices to communicate and also share their data with the controllers or the decision-making centers. IoT platforms also offer the flexibility to select the type of the communication technologies (each having specific

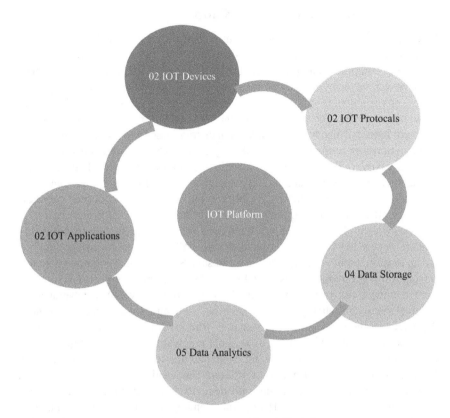

Figure 15.7 Diagram describing the components of an IoT platform.

features) on the basis of the needs of the application. The examples of these technologies include Wi-Fi, Bluetooth, ZigBee, and cellular technology such as LTE-4G and 5G networks. The data storage is the fourth component of the IoT platform, which enables management of collected data from the sensors.

The data collected from the devices is very large, which necessitates planning an efficient data storage that can be in cloud servers or at the edge of an IoT network. The stored data, which is used for analytical purposes, constructs the fifth component of the IoT platforms. The data analytics can be performed off-line after storing the data or it may be in form of real-time analytics. The data analytics are performed for decision making about the operation of the application. Based on the requirement, the data analytics can be performed offline or in real time. In offline analytics, the stored data is first collected and then visualized on premises using visualization tools. In case of real-time analytics, the cloud and edge servers are used to provide visualization, for example, stream analytics.

15.5.2 IoT and energy generation

Automating industrial processes and supervisory control and data acquisition systems became very popular in the power sector in 1990s. By monitoring and controlling equipment and processes, the early stages of IoT started to contribute to the power sector by alleviating the risk of loss of production. Reliability, efficiency, environmental impacts, and maintenance issues are the main challenges of old power plants. The age of equipment in the power sector and poor maintenance problems can also lead to high level of energy losses and unreliability. Assets are sometimes more than 40 years old, very expensive, and sometimes cannot be replaced easily. IoT can contribute to reducing some of these challenges in the management of power plants too. By applying IoT sensors, Internet-connected devices are able to distinguish any failure in the operation or abnormal decrease in energy efficiency and alarming the need for maintenance. This increases reliability as well as efficiency of the system, in addition to reducing the cost of maintenance. A new IoT-based power plant can save $230 million during the lifetime and an existing plant with the same size can save $50 million if equipped with the IoT platform. For reducing fossil fuel use and relying on local energy resources, more countries are promoting residual energy sources. Weather-dependent or variable renewable energy sources, such as wind and solar energy, pose new challenges to the energy system known as "the intermittency challenge." In an energy system with a high share of variable renewable energy, matching generation of energy with demand is a big challenge due to variability of supply and demand, resulting in a mismatch in different time scales. IoT systems offer the flexibility in balancing generation with demand, which in turn can reduce the challenges of deploying variable renewable energy, resulting in higher integration shares of clean energy and less greenhouse (GHG) emissions. In addition, by employing IoT, a more efficient use of energy can be achieved by using machine learning algorithms that help determine an optimal balance of different supply and demand technologies. For instance, the use of artificial intelligence algorithms can balance the power output of a thermal power plant with the sources of in-house power generation, for example, aggregating many small-scale solar photo voltaic panels. Table 15.2 summarizes the applications of IoT in the energy sector, from energy supply regulation and markets.

15.5.3 Smart cities

In the present times, the staggering rate of urbanization as well as overpopulation has brought many global concerns, such as air and water pollution, energy access, and environmental concerns. In this context, one of the main challenges is to provide the cities with clean, affordable, and reliable energy sources. The recent developments in digital technologies enable application of smart, IoT-based solutions for the existing problems in a smart city context. Smart factories, smart homes, power plants, and also the farms in a city can be connected and the

Table 15.2 Applications of IoT in the energy sector (I): regulation, market, and energy supply side

	Application	Sector	Description	Benefits
Regulation and market	Energy democratization	Regulation	Providing access to the grid for many small end users for peer-to-peer electricity trade and choosing the supplier freely	Alleviating the hierarchy in the energy supply chain, market power, and centralized supply; liquifying the energy market and reducing the prices for consumers; and creating awareness on energy use and efficiency.
	Aggregation of small prosumers (virtual power plants)	Energy market	Aggregating load and generation of a group of end users to offer to electricity, balancing, or reserve markets	Mobilizing small loads to participate in competitive markets; helping the grid by reducing load in peak times; Hedging the risk of high electricity bills at peak hours; and improving flexibility of the grid and reducing the need for balancing assets. Offering profitability to consumers.
Energy supply	Preventive maintenance	Upstream oil and gas industry/utility companies	Fault, leakage, and fatigue monitoring by analyzing of big data collected through static and mobile sensors or cameras.	Reducing the risk of failure, production loss and maintenance downtime; reducing the cost of O&M; and preventing accidents and increasing safety.
	Fault maintenance	Upstream oil and gas industry/ utility companies	Identifying failures and problems in energy networks and possibly fixing them virtually.	Improving reliability of a service; improving speed in fixing leakage in district heating or failures in electricity grids; and reducing maintenance time and risk of health/safety.
	Energy storage and analytics	Industrial suppliers or utility companies	Analyzing market data and possibilities for activating flexibility options such as energy storage in the system.	Reducing the risk of supply and demand imbalance; increasing profitability in energy trade by optimal use of flexible and storage options; and ensuring an optimal strategy for storage assets.
	Digitalized power generation	Utility companies and system operator	Analyzing big data of and controlling many generation units at different time scales.	Improving security of supply; improving asset usage and management; reducing the cost of provision of backup capacity; accelerating the response to the loss of load; and reducing the risk of blackout.

data about their energy consumption in different hours of the day can be gathered. If it is found that a section, such as residential areas, consumes the most energy in the afternoon, then automatically energy devoted to other sections, like factories, can be minimized to balance the whole system at a minimum cost and risk of congestion or blackout. In a smart city, different processes, that is, information transmission and communication, intelligent identification, location determination, tracing, monitoring, pollution control, and identity management can be managed perfectly by the use of IoT technology. IoT technologies can help to monitor every object in a city. Buildings, urban infrastructure, transport, energy networks, and utilities could be connected to sensors in an IoT architecture. These connections can ensure an energy-efficient smart city by constant monitoring of data gathered from sensors. For example, by monitoring vehicles with IoT, streetlights can be controlled for the optimal use of energy. Also, the authorities can have access to the gathered information and can make more informed decisions on transportation choices and their energy demand.

15.5.4 Smart grid

Smart grids are modern grids deploying the most secure and dependable ICT technology to control and optimize energy generation, transmission, and distribution (T&D) grids, and their end usage. By connecting many smart meters, a smart grid develops a multidirectional flow of information, which can be used for optimal management of the system and efficient energy distribution. The application of smart grid can be highlighted in different subsectors of the energy system individually such as energy generation, buildings, or transportation, or they can be considered altogether. Also IoT can be applied in isolated and microgrids for some islands or organizations, especially when energy is required every single moment with no exception like in databases. In such systems, all the assets connected to the grid can interact with each other. Also, the data on energy demand of any asset is accessible. This interaction can assure the perfect management of the energy distribution whenever, wherever, and everywhere needed. In terms of collaborative impact of smart grids, as it is shown in Figure 15.8, in a smart city equipped with IoT-based smart grids, different sections of the city can be connected together.

15.5.5 Smart building

The energy consumption in cities can be divided into different parts: residential buildings, and commercial, including shops, offices, and schools, and transport. The domestic energy consumption in the residential sector includes lighting, equipment (appliances), domestic hot water, cooking, refrigerating, heating, ventilation, and air conditioning (HVAC) (Figure 15.9). HVAC energy consumption typically accounts for half of energy consumption in buildings. Hence, the management of HVAC systems is important in reducing electricity consumption. With the advancement of technology in the industry, IoT devices can play an important

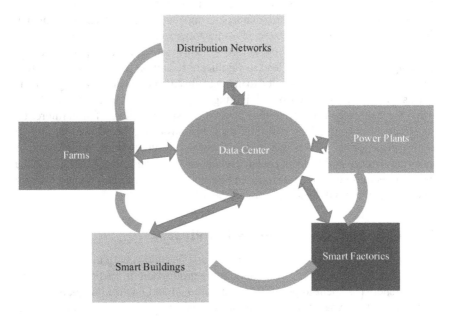

Figure 15.8 A centralized data connectivity in a smart city concept.

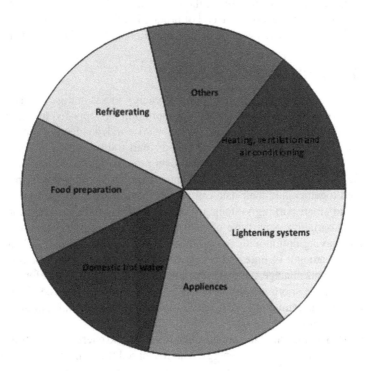

Figure 15.9 Share of residential energy consumption.

role to control the energy losses in HVAC systems; for example, by locating some wireless thermostats based on occupancy, unoccupied places can be realized. Once an unoccupied zone is detected, some actions can be taken to lower energy consumption. For the time being, HVAC systems can reduce the operation in the unoccupied zone, which will lead to significant reduction in energy consumption and losses. IoT can also be applied to manage the energy losses of lighting systems like as through applying IoT-based lighting systems, the customers will be alerted when the energy consumption goes beyond the standard level. Furthermore, by an efficient analysis of the real-time data, load from high-peak will be shifted to low-peak levels. This makes a significant contribution to the optimal use of electrical energy and reducing related effect of GHG emissions. Using IoT, the demand response will be more agile and flexible, and the monitoring and demand side management will become more efficient.

15.5.6 Smart use of energy in industry

IoT can be used to design a fully connected and flexible system in the industry to reduce energy consumption while optimizing production. In traditional factories, a lot of energy is spent to produce the end product and control the quality of the end product too. Apart from it, monitoring every single process requires human resources to be involved. Even so, using an agile and flexible system in smart factories helps to recognize failures at the same time rather than recognizing them by monitoring the products at the end of production line. Hence, a suitable action can be taken promptly to avert wasteful production and associated waste energy.

15.5.7 Intelligent transportation

One of the major causes of air pollution and energy losses in big cities is overuse of private vehicles in place of public transportation. As opposed to a traditional transportation system where each system works independently, applying IoT technologies in transportation, the so-called smart transportation offers a global management system. Also, the real-time data processing plays a significant role in traffic management. All the components of the transportation system can be connected together, and their data can be processed together as well. Congestion control and smart parking systems using online maps are a few applications of smart transportation. Smart use of transportation enables passengers to select a more cost-saving option with shorter distance and the fastest route, which saves a significant amount of time and energy. Citizens will be able to determine their arrival time and manage their all schedule more efficiently. Therefore, time of city trips will be shortened, and the energy losses will be reduced significantly as well. This can remarkably reduce CO_2 emissions and other air polluting gases from transportation.

Table 15.3 summarizes the applications of IoT in the energy sector, from smart energy grids to the end use of energy. The IoT-based digitalization transforms an

Table 15.3 Applications of IoT in the energy sector (2): energy grids and demand side

Application	Sector	Description	Benefit
Transmission and distribution (T&D) grid of EVs.			
Smart grids	Electric grid management	A platform for operating the grid using big data and ICT technologies as opposed to traditional grids	Improving energy efficiency and integration of distributed generation and load; improving security of supply; and reducing the need for backup supply capacity and costs
Network management	Electric grid operation & management	Using big data at different points of the grid to manage the grid more optimally.	Identifying weak points and reinforcing the grid accordingly and reducing the risk of blackout
Integrated control of electric vehicle fleet (EV)	Electric grid operation & management	Analyzing data of charging stations and charge/discharge cycles of EVs.	Improving the response to charging demand at peak times; analyzing and forecasting the impact of EVs on load; and identifying areas for installing new charging stations and reinforcement of the distribution grid.
Control and management of vehicle to grid (V2G)	Electric grid operation & management	Analyzing load and charge/discharge pattern of EVs to for supporting the grid when needed	Improving the flexibility of the system by activating EVs in supplying the grid with electricity; Reducing the need for backup capacity during peak hours Control and management of EV fleet to offer optimal interaction between the grid and EVS.
Microgrids	Electricity grid	Platforms for managing a grid independent from the central grid.	Improving security of supply; creating interoperability and flexibility between microgrids and the main grid; and offering stable electricity prices for the consumers connected to the microgrid.
Control and management of the district heating (DH) network	DH network	Analyzing big data of the temperature and load in the network and connected consumers	Improving the efficiency of the grid in meeting demand; reducing the temperature of hot water supply and saving energy when possible; and identifying grid points with the need for reinforcement.

(continued)

Table 15.3 Cont.

	Application	Sector	Description	Benefit
Demand side	Demand response	Residential/ commercial & industry	Central control (i.e., by shedding, shifting, or levelling)	Reducing demand at peak time, which itself reduces the grid congestion
	Demand response (demand-side management)	Residential/ commercial & industry	Central control (i.e., by shedding, shifting, or leveling; load of many consumers by analyzing the load and operation of appliances)	Reducing demand at peak time, which itself reduces the grid congestion; reducing consumer electricity bills; and reducing the need for investment in grid backup capacity.
	Advanced metering infrastructure	End users	Using sensors and devices to collect and analyze the load and temperature data in a consumer site.	Having access to detailed load variations in different time scale; identifying areas for improving energy efficiency (for example overly air-conditioned rooms or extra lights when there is no occupants); and reducing the cost of energy use.
	Battery energy management	End users	Data analytics for activating battery at the most suitable time	Optimal strategy for charge/discharge of battery in different time scale; improving energy efficiency and helping the grid at peak times; and reducing the cost of energy use.
	Smart buildings	End users	Centralized and remote control of appliances and devices.	Improving comfort by optimal control of appliances and HVAC systems; reducing manual intervention, saving time and energy; increasing knowledge on energy use and environmental impact; improving readiness for joining a smart grid or virtual power plant; and improved integration of distributed generation and storage systems.

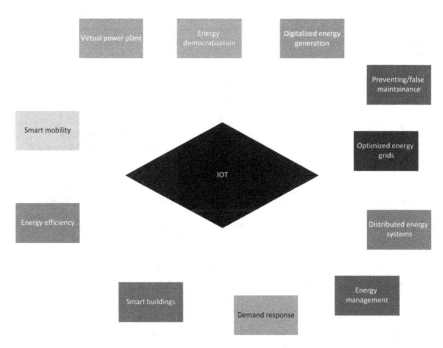

Figure 15.10 Applications of IoT in an integrated smart energy system.

energy system from a unidirectional direction, that is, from generation through energy grids to consumers and to an integrated energy system. Different parts of such an integrated smart energy system are depicted in Figure 15.10.

15.5.8 Challenges of applying IoT

Besides all the benefits of IoT for energy saving, the deploying IoT in the energy sector represents a challenge that has to be addressed. This part addresses the challenges and existing solutions for applying IoT-based energy systems. In addition, in Table 15.4, all the challenges are summarized, which stress the need for current solutions of using IoT in the energy sector.

15.6 CONCLUSION

Energy systems are on the threshold of a new transition based on modern technologies currently available. Large-scale deployment of residual energy in distributed energy systems and the requirement for efficient use of energy calls for system-wide, integrated approaches to minimize the socioeconomic and environmental impacts of energy systems. In this context, modern technologies such as IoT can

Table 15.4 Challenges and current solutions of using IoT in the energy sector

Challenge	Issue	Example Solution	Benefit
Architecture design	Providing a reliable end-to-end connection	Using heterogeneous reference architectures	Interconnecting things and people
	Diverse technologies	Applying open standard	Scalability
Integration of IoT with subsystems	IoT data management	Designing co-simulation models	Real-time data among devices and subsystems
	Merging IoT with existing systems	Modelling integrated energy systems	Reduction in cost of maintenance
Standardization	Massive deployment of IoT devices	Defining a system of systems	Consistency among various IoT devices
	Inconsistency among IoT devices	Open information models and protocols	Covering various technologies
Energy consumption	Transmission of high data rate	Designing efficient communication protocols	Saving energy
	Efficient energy consumption	Distributed computing techniques	Saving energy
IoT Security	Threats and cyber-attacks	Encryption schemes, distributed control systems	Improved security
User privacy	Maintaining users' personal information	Asking for users' permission	Enables better decision-making

help the energy sector transform from a central, hierarchical supply chain to a decentralized, smart, and optimized system. In this chapter, two different types of energy-efficient architectures were discussed and the role of IoT in the energy sector was reviewed in general and in the context of smart grids, in particular. The classification of different use cases of IoT in each section of the energy supply chain, from generation through energy grids to end use sectors were presented. The advantages of IoT-based energy management systems in increasing energy efficiency and integrating renewable energy were discussed and the findings are summarized as well. Different components of an IoT system were discussed, including enabling communication and sensor technologies with respect to their application in the energy sector, for example, sensors of temperature, humidity, light, speed, passive infrared, and proximity. A discussion of the cloud computing and data analytic platforms, which are data analysis and visualization tools that can be employed for different smart applications in the energy sector, from buildings to smart cities, was included. The application of IoT in the energy supply chain under different levels, including smart cities, smart grids, smart buildings, and intelligent transportation was extensively discussed. Some of the challenges of applying IoT in the energy sector, including challenge of identifying objects, big data management, connectivity issues and uncertainty, integration of subsystems, security and privacy, energy requirements of IoT systems, standardization, and architectural design were also explained.

Chapter 16

Exploring robotics technology for health care applications

Swagat Kumar Samantaray and
Shasanka Sekhar Rout

CONTENTS

16.1 INTRODUCTION

In this current pandemic situation, the health care industry is facing a considerable amount of pressure and challenges for improving efficiency, accessibility and most importantly safety of the frontline health care workers. This has led to the accelerated adoption of robotics technology to improve the quality of health care services. The major capacity of the robot is to work like humans, and also work step by step according to the scope of its usage. Robots are used in every field of science and in hospitals, where they are strengthening the existing health care facilities. Robots can be used to support people with sensory and cognitive disabilities in their rehabilitation, provide assistance to caregivers and aid the clinical workforce. Also health care robotics is one of the developing and innovative fields of research that ensures safety of the patients. The technological advancement in machine learning (ML) techniques and Internet of Things (IoT) was intended to be used with robotics technology for health care applications [1, 2]. A number of robots have been developed for monitoring, assistance, care and control applications to provide high-quality patient care in a safe and secure

DOI: 10.1201/9781003335801-16

environment. Collaborative robots have been adopted by the health care industry to automate the medical processes.

16.2 BENEFITS OF ROBOTICS IN HEALTH CARE

Patient care is one of the most primary tasks of medical robots with the support of minimally invasive and a customized patient monitoring system for chronic diseases. It is mostly used by the elderly patients for interaction as well as care. The robot has the capability of providing required guidance in accordance with the situation faced by the elderly people.

Another category of robots is used for operational efficiencies and keeps track of inventory and medical supplies with a suitable algorithm. Service robots in the medical industry are used for reducing the physical demands on human labour and ensure more reliable processes with streamlined routines. Sanitation is of paramount importance in the current pandemic situation; hence the cleaning robot is used to reduce hospital-acquired infections.

Another helping robot or social robot is used for lifting, such as moving beds or patients in hospitals to reduce the physical strain of health care workers. There are specialized intelligent robots and wearable systems for rehabilitation to support and improve the functioning of the health care system.

Medical robot is considered important according to its application. The robots offer various solutions for a particular group of patients who needs special attention. Depending on the application, there are specifically five areas where medical robots can be used:

- Surgical robot
- Prosthetics robot
- Motor control robot for treatment
- Assisted robot for therapy
- Patient monitoring robot

16.3 CLASSIFICATION OF ROBOTS UTILIZED IN HEALTH CARE

The capabilities of robots that make it suitable for application in health care are greater accuracy, precise diagnosis, intervention from remote location and improving the abilities of health care workers. The current pandemic situation provides an opportunity to adopt different robotics technology at a larger scale than before. In critical surgery, by using medical robots, the operation time and risk are reduced. In many countries, most of the tasks related to the cardiac surgery is performed with the help of advanced robots. The capability of the robot extends to provide rehabilitation with a smart prosthesis for different parts of the human body., In terms of patient care, care and monitoring of the elderly people

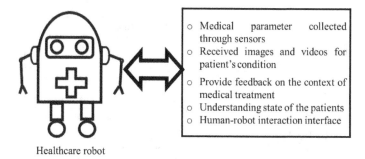

o Medical parameter collected through sensors
o Received images and videos for patient's condition
o Provide feedback on the context of medical treatment
o Understanding state of the patients
o Human-robot interaction interface

Healthcare robot

Figure 16.1 Capabilities of robotics in health care.

can be taken care by social robots. Robotics technology and its capabilities for serving health care are shown in Figure 16.1.

The robot can monitor the medical parameters through suitable sensors. The patient's data in the form of image or video can be communicated to the doctors. It is also able to provide feedback concerned with medical treatment. With an appropriate interfacing media, it can enable the interaction between the patients and doctors [3].

Depending upon the application in health care, the robots can be divided into surgical robots, wearable robots, rehabilitation robots, care robots and collaborative robots. The surgical robots are used in surgical procedures to overcome the limitation of the existing invasive surgical operating procedures and enhance the capability of doctors performing the surgery. The surgeons need to operate the instrumented surgical robot either through direct telemetry or through the computer-controlled robotic arm. The wearable robots are now moving from laboratory research environment to real-time implementation with specific characteristics such as kinematics and shape adopted by end-users. Different human limbs can be restored with suitable prosthetic robots. Most rehabilitation robots are used in the recovery process of disabled patients for standing, balancing and gait. This type of robot must work with patients to monitor their progress. It is designed with the application of techniques that determine the adaptability level of the patient [4].

Service robots have been used as assistants, guides, and companions. Also, it can be applied in health care facilities for delivering goods, medicine and used as a companion for an ageing population. With the current technology development, robots are venturing into the health care industry as medical assistants. The collaborative robots are used for monitoring the patient's biological parameters and alerting the doctors/nurses and also simultaneously monitor a number of patients. The overview of the robotics application in health care is shown in Figure 16.2.

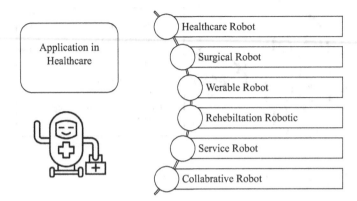

Figure 16.2 Robotics application in health care.

In case of medical robots, there are certain important factors like motion, automatic operation, and intelligence. In motion, the actuated mechanism of the motors is programmed with more than one axis in a moving environment. The automation of robotic device reveals the normal operation or any kind of fault in the system. The intelligence built with expert knowledge allows the users to perform the task in an efficient way.

16.3.1 Health care robot

The overall architecture of a health care robot system consists of different subsystems, which include robots for reception, a robot for nurse assistant and a server for medical data. The robot at the reception can be used for interaction with patients or visitors, which depends on the implemented algorithm for particular hospitals. Another robot can work in the health care facility to get the information and health data of patients and display it on a screen.

The assistant robot can be used for monitoring critical patient information such as blood pressure, pulse rate and blood oxygen level. The robot is helpful for doctors and nurses to obtain the screening data of patients remotely. The system overview is shown in Figure 16.3.

The robot, which is able to manage the communication among the components, includes various software platforms such as robot operating system (ROS), OpenRTM with the external sensor system [5]. The robot will be able to manage the status of each connected framework. For operating in diverse environments, the sensor management unit will manage the entire external sensor units and assist the nurses/doctors. It collects the health information of patients and provides necessary feedback or alerts the concerned health care worker in case of an emergency. Figure 16.4 shows the system architecture of care robot. It consists

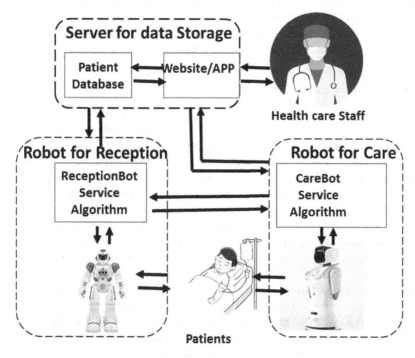

Figure 16.3 Overview of the assistant health care robot.

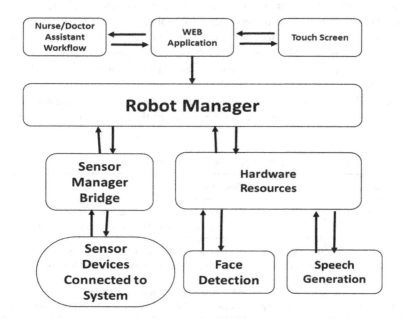

Figure 16.4 Architecture of a care robot.

of different parts, which include the sensor manager and hardware resources. When a patient interacts with the robot to measure different parameters in health care, it sends request to sensor management unit using predefined mobile or web application programming interface (API) [6].

16.3.2 Surgical robot

Most of the robots used for surgical purposes is very minimally invasive with high accuracy and precision as compared to operation by the humans. For remote operation, robot uses the teleportation system. The Da Vinci robot is one type of surgical robot where doctors operate and control through robotics arm from the computer console by looking at the surgical site via a sophisticated magnified camera. It is an advanced system that works on the principle of controlling the master and slave.

The Da Vinci system is an advanced tele-manipulation robot developed by Intuitive Surgical, Inc. for laparoscopic surgery as a minimal invasive option [7]. As it works in a master–slave control mode, it has two units; the first one is for surgeons with a console and a display unit and second unit has slave manipulators for holding surgical instruments and a camera. The robotics system provides a convincing environment to medical experts, which provides an interfacing to hand gesture of doctors to the movement of operating tips inside the patients [8–10].

It is a complex instrument with number of robotic arms and high-power camera for capturing the operated area. The communication between the doctors and patients become efficient and easy to access using the tele-robot system, which enables people who stay in remote areas and having poor medical facilities access to medical facilities. By using the robotic technology, services can be extended to more number of people. Most of the developing countries are using the existing possibilities and are able to make the robotic surgery more effective and afford-able. Most commonly used surgical robots and their applications are listed in Table 16.1.

Generally, the robotic surgery is performed based on the human parts and avenues. Normally, they are of two types based on the human parts: neurosurgery

Table 16.1 Different surgical robots and its application

Robot Name	Make	Description
Da Vinci	Intuitive	For holding different tools, a robot arm is used with a high-definition 3D camera as patient view.
Mako	Stryker	Joint replacement surgery
Navio	Smith & Nephew	3-D imaging techniques for robot assistant
Auris	Auris health	The flexible robotic arms are used for endoscopy process
ROSA	ROSA	360° Imaging technique for knee
KUKA LBR	KUKA, Augsburg	To carry out complex surgical procedure with robot

and cardiac surgery. In neurosurgery, the robot deals with actions relating to the nervous system, which includes the nerves, brain and the spinal cord. During the surgery, doctors use the robotic arm having six degrees of freedom and use reflectors that are connected near the operating parts with a suitable high-power camera [11].

In case of robotic cardiac surgery, specially designed instruments are used. It makes use of incisions or port, which is embedded with a camera along with surgical instruments. Apart from this, a motion sensor is provided to give a proper control to the surgeon. An arrangement of foot control has been given for zooming the camera [12].

16.3.3 Rehabilitation robot

Rehabilitation robot is one type of automatic instrument designed for the progressive movement in patients who have impaired physical function of a body part. It is designed based on the adaptability level of patients. Mostly the rehabilitation robots are of two different kinds. The first one is an assistive robot, which is a replacement of the lost limb movement and the second kind is a therapy robot, which is also named as a rehabilitator. In assistive robots, collaborative effort between the machine and human limbs are actuated through some actuators and they behave like the muscles, which resembles actual biological locomotion [13,14].

To design the prosthetic limb, sophisticated motors, sensors to record the muscle signals, artificial muscles that are able to mimic the property of actual muscles are employed. This robot is used in the rehabilitation process to assist the physiotherapists. During the process of physical therapy, the patients will perform some sets of prescribed exercises under the supervision of a professionally trained person. On the other hand, it is laborious, costly and challenging for patients [15]. Hence, a robot helps to perform the repetitive task and provides precise assistive movement for the patients.

The robot using multiple degrees of freedom can be used for the rehabilitation process. To enhance the quality of life for people with physical disorders and elderly people, regular exercises are required, which can be fulfilled by a rehabilitation robot. The human–machine interaction (HMI) is one of the essential parts of this type of robotics systems that can stimulate the recovery effect of the patients under treatment. A number of rehabilitation robots have developed to help patients to recover quickly [16,17]. For rehabilitation process, the robots provide training and tasks to guide the patients as per the scheduled algorithm. It can also provide repetitive physical therapy to the patients. The rehabilitation robot used for training is presented in Figure 16.5.

The electrical activity produced by nerve cells is in the mill volt range. When the brain does some conceptual work of creating a movement or motion in mind, then these nerve cells produce a number of signals, which can be transmitted through suitable electrodes connected at the scalp of the patients. The pattern of

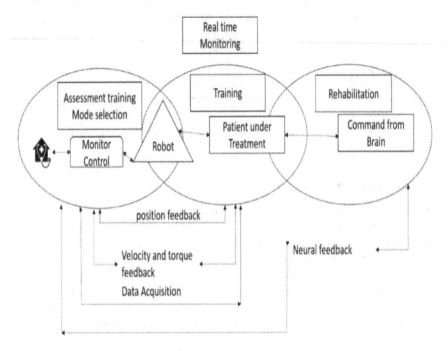

Figure 16.5 Working of rehabilitation robot system.

the electroencephalograph (EEG) signal has some characteristics and processing this rhythmic pattern provides a signal of human's intention. Then these signals are then converted into a suitable control command to help the supervisor of the external equipment that is connected and communicated to the outside world. This human–machine interface principle is shown in Figure 16.6.

The different wearable and rehabilitation robotic technology with its actuation strategy is tabulated in Table 16.2.

16.3.4 Wearable robot

The design of the wearable robot for a particular type of disorder and prescribing the suitable rehabilitation is one of the crucial tasks. The robot must be designed based on the algorithm and control strategy to adapt to various phenomena. The robotics glove is designed using an electro pneumatic system with a controller, which provides individual control to each actuator. A wearable glove has been designed to implement the control strategy. It is a five-finger device with soft plastic motors embedded in the glove. A microcontroller with a suitable supply voltage regulator is used as a control system. A pneumatic control sensor is used to regulate the air pressure for each actuator. To regulate the air pressure (P) and

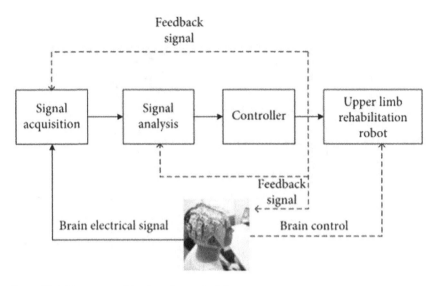

Figure 16.6 Human–machine interface principle.

Table 16.2 Rehabilitation and assistance robots and its technology

Rehabilitation system name	Type	DOF	Actuation scheme	Sensor
Exo-Glove Poly	Soft glove	2	Electric; Two DC motors (antagonistic arrangement)	Handheld switch
Xiloyannis	Soft exosuit	2	Electric system with two DC motors and to control index a tendon-driving unit, middle and thumb with single motor	Flex sensor
Soft glove	Soft glove	1	Electric; SMA; SMA actuator with flexible potentiometer	Flexible potentiometer
TU Berlin Finger Exoskeleton,	Rigid orthosis	4	Electric; DC motors; with joint angle actuators	Joint angles
hand mentor, kinematic muscles	Rigid orthosis soft actuator	1	Pneumatic	Angle and torque sensor
SEM Glove	Soft glove	3	Electric; DC motors; Bowden cable	capacitive sensors and force-sensing

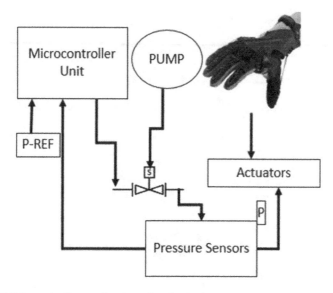

Figure 16.7 Schematic diagram for the soft robotic glove.

to track the desired value of pressure (P-REF), the microcontroller uses the pulse width modulation (PWM) technique. By the PWM, it controls the activation and deactivation of the valves based on the pressure sensor data. The block diagram of robotics glove is shown in Figure 16.7.

Another type of robotic glove is designed using a controller with potentiometers, accelerometer sensor and flex sensor. The robot arm will get the motion from servo motors, which are connected through a servo motor driver. The motors control the movement of each joint in the robotic arm. The placement of the servo motor decides the number of degrees of freedom and load on the arm depends on the motor specification [18,19].

The finger movement mechanism of robot arm is related to the finger movement of the human hand. The movement of the arm can be controlled via a controller with a sensor-based glove [20,21]. The control setup of robot arm is shown in Figure 16.8 with all modules and its interconnection [4].

The sensors mounted on gloves are one type of variable resistive type. The sensors provide the voltage with respect to movement of the hand and give a signal to the controllers for controlling the movement of serving motors. One of the most commonly used resistive sensors is flex sensors, which are based on carbon elements. The sensor works on the principle of variable resistance, which measures the amount of deflection or bending. Here the bend sensors are used for each finger and corresponding movement can be recorded by a suitable controller. The design of a robotic glove is shown in Figure 16.9. The PWM technique is

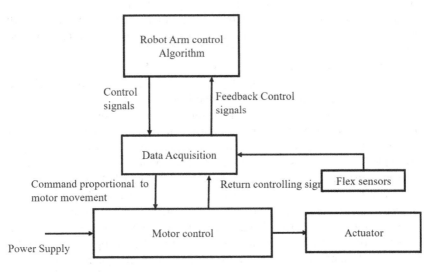

Figure 16.8 Control setup of robot arm.

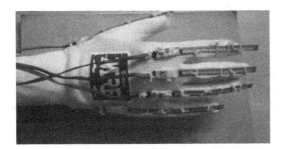

Figure 16.9 Design of five-fingered robotic globe.

used for controlling the servo motors, with different duty cycles. The different duty cycles are maintained by manipulating time and by using suitable PWM signals [22].

The virtual reality (VR) head mounted display is one major technologies employed in wearable robotics. During the pandemic situation, it is adopted for public health and helps medical professionals as it provides a better platform for tele-robot control. The VR is also able to bring the doctors together and allow physicians to safely provide expert advice to patients [23].

16.3.5 Robots for care

The care robots are designed to help people and are also capable of measuring the behaviour, cognitive state of patients and offers some kind of therapeutic effects. The characteristics of the care robot depends on the way it is used for providing assistance for nurses and doctors or support for patients, children or elderly people. A number of robots with necessary hardware and explicit operation are available for use by the elderly people.

Nao Robot is one type of humanoid robot having 25 degrees of freedom (DOF) out of which 11 of them are in the legs and pelvis joint of the robot. The other DOFs are for each hand, two each for the head and hips. It also has different sensor units such as RGB cameras, tactile sensors, sonar sensor for range finding, microphones for recording, infrared emitters and receivers for detection and pressure sensors to monitor the amount of load pressure. The robot uses the open-source platform with the support of programming language like C++ and Python. Nao can be used as an assistive robot to train elderly people [24,25]. The robot can have a number of programmed therapeutic behaviours. Nao is also used for educational purposes for rehabilitation training of children with physical disabilities.

The robot is used for providing assistance to elderly patients and children. The robot can provide physical, cognitive and social support to people. The care robot has a huge demand in recent times, especially from elderly people. A care robot also conducts various activities based on the physical emotions and provides assistance or feedback to the doctors, support to the caregivers and also used for social interaction.

PHAROS is an interactive robot specially designed for elderly people to monitor their daily physical activity, either in hospital or at home [26]. This type of robot has the capability of vision and deep learning (DL) techniques are implemented. Generally, the camera unit of the robot records the activity of patients or the person and provides data that will be fed to the system to recognize the activity. Based on the activity, a sequence of tasks has been prescribed by the robot for individual users. The beauty of the system is that it evaluates the performance of the patients without the presence of a therapist.

Recently, socially assistive robots (SARs) have been designed, which operate automatically and does not require any manipulation or handling by a human operator [27]. The SARs can provide a positive impact on patients and reduce their stress level, improve their moods and used for better communication with the doctors and the social environment. In this category, nurse robot is used in hospitals to assist doctors resembling the work of nurses. Mostly the nurse robots are used by ageing family members to take care in a home environment. Nurse robots are offering day and night services in an efficient way.

16.4 CONCLUSION

The robotics technology and its potential for developing various applications offer tremendous assistance to the health care industry. Employing modern technology with capabilities of smart sensors for health care makes the device accessible through IoT, which is one of the key factors for the success of robotics technology. The chapter provided an insight into the various robotics technologies and its demand from health care, which can benefit all the stockholders. In this chapter, an overview of surgical robotics, wearable robotics and technology to provide a social support have been covered from the medical community point of view. The wearable robot used for a specific disorder and in which suitable rehabilitation is built in was discussed by using a suitable procedure. The care robots are used for helping elderly people and are also capable of measuring the routine behaviour, cognitive status and offer therapeutic effects for improvement.

References

[1] Zemiti, Nabil, Tobias Ortmaier, and Guillaume Morel. "A new robot for force control in minimally invasive surgery." International Conference on Intelligent Robots and Systems (IROS) (IEEE Cat. No. 04CH37566) 4 (2004): 3643–3648.

[2] Pugin, F., P. Bucher, and Philippe Morel. "History of robotic surgery: from AESOP® and ZEUS® to da Vinci®." *Journal of Visceral Surgery* 5.148 (2011): e3–e8.

[3] Okamura, Allison M., Maja J. Matarić, and Henrik I. Christensen. "Medical and health-care robotics." *IEEE Robotics & Automation Magazine* 17.3 (2010): 26–37.

[4] Qian, Zhiqin, and Zhuming Bi. "Recent development of rehabilitation robots." *Advances in Mechanical Engineering* 7.2 (2015): 563062.

[5] Ahn, Ho Seok, Min Ho Lee, and Bruce A. MacDonald. "Healthcare robot systems for a hospital environment: CareBot and ReceptionBot." 24th IEEE International Symposium on Robot and Human Interactive Communication (RO-MAN). IEEE, 2015.

[6] Fan, Jing, et al. "A robotic coach architecture for elder care (ROCARE) based on multi-user engagement models." *IEEE Transactions on Neural Systems and Rehabilitation Engineering* 25.8 (2016): 1153–1163.

[7] Hubens, G., et al. "A performance study comparing manual and robotically assisted laparoscopic surgery using the da Vinci system." *Surgical Endoscopy and Other Interventional Techniques* 17.10 (2003): 1595–1599.

[8] P. Maciejasz, J. Eschweiler, G.-H. Kurt, J.-T. Arne, S. Leonhardt, K. Gerlach-Hahn, et al., A survey on robotic devices for upper limb rehabilitation, *Journal of Neuroengineering Rehabilitation* 11.1 (2014) 1–29.

[9] Pandey, A.K.; Gelin, R. "A mass-produced sociable humanoid robot: pepper: the first machine of its kind." *IEEE Robotics and Automation Magazine* 25 (2018) 40–48.

[10] Freschi, Cinzia, et al. "Technical review of the da Vinci surgical telemanipulator." *The International Journal of Medical Robotics and Computer Assisted Surgery* 9.4 (2013): 396–406.

[11] Khanna, Omaditya, et al. "The path to surgical robotics in neurosurgery." *Operative Neurosurgery* 20.6 (2021): 514–520.

[12] Diodato, Michael D., and Ralph J. Damiano. "Robotic cardiac surgery: overview." *Surgical Clinics* 83.6 (2003): 1351–1367.

[13] Tejima, Noriyuki. "Rehabilitation robotics: a review." *Advanced Robotics* 14.7 (2001): 551–564.

[14] Kumar, S. S., Ahmed, S. T., Vigneshwaran, P., Sandeep, H. and Singh, H. M. Two phase cluster validation approach towards measuring cluster quality in unstructured and structured numerical datasets. *Journal of Ambient Intelligence and Humanized Computing* 12.7 (2021): 7581–7594.

[15] Pezzin, Liliana E., et al. "Use and satisfaction with prosthetic limb devices and related services." *Archives of Physical Medicine and Rehabilitation* 85.5 (2004): 723–729.

[16] Vallery, Heike, et al. "Compliant actuation of rehabilitation robots." *IEEE Robotics & Automation Magazine* 15.3 (2008): 60–69.

[17] Bessler, Jule, et al. "Safety assessment of rehabilitation robots: A review identifying safety skills and current knowledge gaps." *Frontiers in Robotics and AI* 8 (2021): 33.

[18] Periasamy, K., Periasamy, S., Velayutham, S., Zhang, Z., Ahmed, S. T., and Jayapalan, A. *A Proactive Model to Predict Osteoporosis: An Artificial Immune System Approach.* Expert Systems, 2021.

[19] Sathiyamoorthi, V., Ilavarasi, A. K., Murugeswari, K., Ahmed, S. T., Devi, B. A., and Kalipindi, M. "A deep convolutional neural network based computer aided diagnosis system for the prediction of Alzheimer's disease in MRI images." *Measurement,* 171 (2021): 108838.

[20] O'Neill, Ciarán T., et al. "A soft wearable robot for the shoulder: Design, characterization, and preliminary testing." 2017 International Conference on Rehabilitation Robotics (ICORR). IEEE, 2017.

[21] Gordleeva, Susanna Yu, et al. "Real-time EEG–EMG human–machine interface-based control system for a lower-limb exoskeleton." *IEEE Access* 8 (2020): 84070–84081.

[22] Ahmed, S. T., Sandhya, M., and Sankar, S. A dynamic MooM dataset processing under TelMED protocol design for QoS improvisation of telemedicine environment. *Journal of Medical Systems* 43.8 (2019): 1–12.

[23] Javaid, Mohd, and Abid Haleem. "Virtual reality applications toward medical field." *Clinical Epidemiology and Global Health* 8.2 (2020): 600–605.

[24] Khosla, Rajiv, Khanh Nguyen, and Mei-Tai Chu. "Human robot engagement and acceptability in residential aged care." *International Journal of Human–Computer Interaction* 33.6 (2017): 510–522.

[25] Khaksar, Seyed Mohammad Sadegh, et al. "Service innovation using social robot to reduce social vulnerability among older people in residential care facilities." *Technological Forecasting and Social Change* 113 (2016): 438–453.

[26] Costa, Angelo, et al. "PHAROS – Physical assistant robot system." *Sensors* 18.8 (2018): 2633.

[27] Bemelmans, Roger, et al. "Socially assistive robots in elderly care: a systematic review into effects and effectiveness." *Journal of the American Medical Directors Association* 13.2 (2012): 114–120.

Chapter 17

Design principles, modernization and techniques in artificial intelligence for IoT

Advanced technologies, development and challenges

P. Radhika Ravi, S. Sarumathi, and Ravi Ramaswamy

CONTENTS

17.1 INTRODUCTION

17.1.1 Internet of Things

IoT is getting smarter with loads of devices and applications that are connected to the internet to collect, process and share data. The IoT is making the world around us faster, smarter, precise and responsive, integrating the physical and the virtual digital universe.

"Universal spending on the IoT has an increase of 15.4% over the $646 billion spent in 2018", according to IDC and likely to pass the $1 trillion mark in 2022.

IoT is an organization of well-correlated set of sensors, devices, physical objects, actuators, virtually connected objects, services, application platforms and people and integrated network. Each object has specific unique identity and hence can transfer data collaboratively or independently.

17.1.2 Artificial intelligence

AI is the computerized model that enables data analytics, hypotheses and decision making from the data collected by the connected devices. In practice, IoT collects the data and the allied artificial intelligence processes this data in order to get required results. Analytics and AI apply machine learning algorithms that are used to process the data collected by the IoT devices. Using these reports and quick references, an organization can identify, discern and understand patterns and make more informed decisions.

IoT empowered with AI engenders intelligent machines that simulate smart behaviour and supports in decision making with minor or no human interference. IoT and artificial intelligence offers wide range of benefits like proactive intervention, personalized experience, and intelligent automation.

Industry 4.0 is the current revolution with some of the emerging fields as building blocks, viz. artificial intelligence (AI), augmented reality, IoT, machine learning, robotics, block chain, nanotechnology, quantum computing, biotechnology and human augmentation.

17.2 THE INDUSTRIAL REVOLUTION

The fourth industrial revolution (4IR or Industry 4.0) is a continuous progressive automation of traditional manufacturing and industrial processes and practices using advance smart technology. Large and wide-scale machine-to-machine communication (M2M), AI and the Internet of Things (IoT) are harmonized and integrated for increased automation, enhanced self-monitoring systems, communication and production of smart applications and machines that can analyse, scrutinize and diagnose issues without much human intervention. Here is the short history of the industrial revolution.

17.2.1 The first industrial revolution

The first industrial revolution began in the eighteenth century through the use of steam power and mechanization of production. The paraphernalia in yesteryears that produced threads on humble spinning wheels now has the mechanized version, which has achieved several times the volume with same time, effort and with better precision. The use of steam for industrial purposes was the greatest breakthrough for increasing productivity and fecundity. Instead of weaving looms using people, steam engines were deployed for power and speed. Developments such as the steamship came in a century later. The steam-powered locomotive accrued massive changes in industry because the movement of goods and people was easier in a lower timeframe.

17.2.2 The second industrial revolution

The second industrial revolution began in the nineteenth century through the discovery of electricity and assembly line production. Henry Ford (1863–1947) took the idea of mass production from a slaughterhouse in Chicago. The pigs hung from conveyor belts and each butcher performed only a part of the task of butchering the animal. Henry Ford carried over these principles into automobile production and drastically altered it in the process. While before one station assembled an entire automobile, now the vehicles were produced in partial steps on the conveyor belt – significantly faster and at a lower cost.[2]

17.2.3 The third industrial revolution

The third industrial revolution Commenced in the 1970s. The automation was quite fragmented, deploying memory-programmable controls and microcomputers. After the advent of these new engineering technologies and use of electrical and electronics, automation of an entire production process is possible without human interference.

17.2.4 The fourth industrial revolution

The fourth industrial revolution is characterized by the application of information and communication technologies to industry and is also known as "Industry 4.0". The term "Industry 4.0" was used in public for the first time at Hannover Messe in 2011 and subsequently integrated into the German government's high-tech strategy. This builds on the developments made in the third industrial revolution. Production systems that were already automated using computers were networked using the internet. This allowed communication with other facilities and provided data as output about themselves. This is the next step in production automation. The networking of all systems leads to "cyber-physical production systems". In smart factories, with cyber-physical system (CPS), production systems, components and people communicate via a network and production is nearly autonomous.

17.3 INDUSTRY 4.0 EXPLAINED

Industrie 4.0 refers to the intelligent networking of machines and processes for the industry with the help of information and communication technology (Platform Industrie 4.0). It is the information-intensive, information-driven transformation of automated manufacturing processes and control systems in a connected environment.

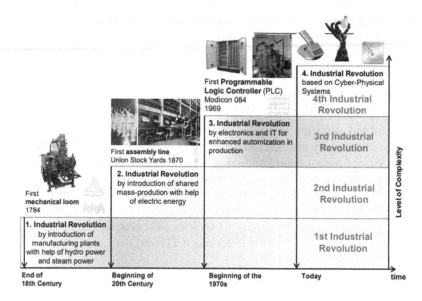

Figure 17.1 The Industry 4.0 revolution.

Industry 4.0's origin is from the German manufacturing powerhouses. Nonetheless, this concept was widely adopted by other industrial nations such as Japan, India, the European Union, China and other Asian countries. The fourth industrial revolution is nothing but confounding of boundaries and integrating the digital, physical and biological worlds. The core and essential element in implementing Industry 4.0 is the advent of cyber-physical systems (CPS).

CPSs are important part of the revolution that is transforming the way humans interact with engineered systems. A CPS is a harmonized system integration of collaborating computers, devices and network elements controlling physical manufacturing and device entity. The CPS are physical and integrated engineered systems, whose operations are planned, coordinated, monitored, controlled and integrated by computers and communication networks. In short, a cyber-physical system is an internet-enabled well harmonized physical entity.

A CPS is composed of an array of varied inhomogeneous elements; therefore, it requires complex models to define each subsystem element and its attributes and behaviour. Harmonizations and dynamic interactions between the sub-systems are orchestrated by an overarching model. The major capabilities of the cyber-physical systems, which form the basis that enables novel capabilities, are product design, prototyping and product development, remote control, services and diagnosis, conditional monitoring, track and trace, structural health and systems health monitoring, planning, innovation capability, agility, real-time applications, proactive and predictive maintenance, and security compliance. Examples of CPS are automated medical monitoring, robotics working safely with humans, detection and surveillance systems, wearable computing/sensing uniforms automatic pilot in avionics and autonomous automobile systems.

17.4 THE BUILDING BLOCKS OF INDUSTRY 4.0

One of the major challenges faced while adopting IIoT is the preference and selection of architecture. Since the concept of IIoT itself is based on connecting things and devices, network architecture plays a critical role. The basic design principle needs are interoperability, virtualization, decentralization, real-time capability, service orientation and modularity. The major elementary blocks of Industrie 4.0 technology are as follows:

- RAMI 4.0 [9]– The Reference Architecture Model Industry 4.0
- Cyber-physical systems
- Internet technology – connectivity and networking
- Manufacturing objects as information carriers
- Holistic approach for safety, security, privacy and knowledge protection

The major technological building blocks of Industry 4.0 are cloud computing, cyber security, augmented reality, IoT and IIoT, big data analytics, artificial intelligence, software-based sequential layering of materials to produce 3D shapes called as additive manufacturing and industrial robotics.

Figure 17.2 further explains the finer segments of the Industry 4.0 pioneering systems.

The ingredients for Industry 4.0			
Instrumented	**Interconnected**	**Inclusive**	**Intelligent**
Data Devices contain sensors, actuators and software that generate data	**Connectivity** An information network connects devices together; gathers and processes the data either at the edge of the network or centrally – selectively	**Context** Industry knowledge, data external to the network (weather etc.) adds context to the data	**Decision Making** Machine learning, predictive analytics and cognitive computing makes sense of the data; decentralized decision making, move towards autonomy

What Industry 4.0 enables		
Design	**Make**	**Use**
Integrate - use of existing products by equipping them with sensors to bring them into the connected environment	**Optimise** - predictive maintenance of production lines optimises uptime and maximises throughput	**Satisfy** – predictive maintenance of products assures optimal usability and availability, optimised supply chains assure availability
Predict - design new products based on utilisation of existing products and market reaction to concepts	**Fulfil** - meet market demands by providing what is most utilised	**Safety** – hazardous tasks and environments are delivered by robots.
Innovate - insight from sensor data can guide equipment usage and new product or service design based on customer use and use across a network	**Extend** - machines will come with intelligence pre-built. The applications for those product-service hybrids will become revenue streams	**Sensory** – new ways for humans to interact digitally with machines through voice, sight, touch and movement.
	Employ - new roles for product and experience designers, application developers, data scientists equipment/network production, implementation and support.	

Figure 17.2 Ingredients of Industry 4.0.

Source: IBM.Com/IOT

17.5 SIX DESIGN PRINCIPLES OF INDUSTRY 4.0

Emerging technologies demonstrate the breadth of applications that make up Industry 4.0. The ecosystem and technology do not operate in isolated factories or assembly lines, rather in environments and devices; technologies connect with other entities, on entire production hierarchies and value chains, consistently throughout the product life cycles. Here are some of the design principles at a high level.

- *Interoperability*: The ability of cyber-physical systems (i.e. work piece carriers, assembly stations and products), humans and smart factories to connect and communicate with each other via the Internet of Things and the Internet of Services.
- *Virtualization*: It is a simulated instance of a smart factory, which is devised by linking data from various sensors that monitor physical processes. Simulation models are deployed to study patterns.
- *Decentralization*: The ability of cyber-physical systems within smart factories to make decisions on their own,
- *Real-time capability*: A dynamic capability which processes collected data, analysis, creates reports and provides the insights immediately.
- *Service orientation*: Managed services offering which provides indented services via the Internet of Services.
- *Modularity*: Easy adaptation and flexibility of smart factories for dynamic requirements of individual modules.

The Industry 4.0 paradigm and framework augments intelligent manufacturing systems to the basic fundamental processes of fabrication and assembly. The integral subsystems and equipment in this environment and ecosystem will be connected online using internet. They are intelligent and capable of making decisions with diverse and varying degrees of autonomy.

17.6 INDUSTRIAL IOT

17.6.1 What is the industrial Internet of Things?

The industrial Internet of Things (IIoT) is the use of smart sensors, actuators, computers and people enabling intelligent industrial operations to enhance manufacturing and industrial floor process.

General Electric (GE) conceived the phrase "Industrial Internet," which describes the industrial transformation in the connected framework of machines, edge computing, cyber systems, analytics, AI, people, cloud, IoT and security.

The smart manufacturing enterprise consists of smart machines, plants and operations, which have higher levels of intelligence embedded from the core. The linked systems are based on open and standard internet and cloud technologies that enable secure access to devices and information / data. In addition, faster and intelligent computation power has been achieved. Machine learning, statistics, and artificial intelligence (AI) algorithms, complex modelling, augmented reality, virtual reality and optimization techniques are becoming essential aspects of the manufacturing floor that makes the factory "smart".

The IIoT is a network of physical objects, systems, platforms and applications that contain internet and embedded technology to communicate and share intelligence with each other, to the external environment and with humans. The adoption of the IIoT is now easy because of improved availability and affordability of sensors, processors, internet and other technologies, which are important to access real-time information.

The life cycles of IIoT applications differ from those of IoT. IoT application life cycles have less corroboration process stages and therefore have faster turnover. On the other hand, the IIoT applications must pass thorough testing, simulation, validation, rectification and verification prior to deployment. The deployment environment of IIoT applications requires integration with areas, such as legacy systems and devices, sensors, simulators, intelligent robotics, big data, AI, analytics, machine learning and augmented reality.

The IIoT technologies will allow for closer integration of production systems and ERP systems, product lifecycle management (PLM) systems, supply chain management and customer relationship management (CRM) systems, which are now operating in silos, harmonizing and facilitating enormous efficiency gain. Subsequently, operations technology (OT) machines will merge with information technology (IT) systems and an integrated automation will evolve, which will be a smooth, noncomplex, and more information-driven architecture.

17.7 ESSENTIAL IIOT TERMS AND CONCEPTS

There are numerous concepts and terms that relate to smart systems, IIoT and Industry 4.0, The foundational words and phrases to know before stepping in and adapting Industry 4.0 solutions for your business are as follows:

- *Enterprise resource planning (ERP)*: Business process management tools that can be used to manage information across an organization integrating various departments such as finance, HR, commerce, supply chain, operations, reporting and manufacturing activities.
- *IoT*: The Internet of Things is a concept that refers to a system of related and integrated computing devices, connections between physical objects like sensors microcontrollers and actuators or machines mechanical and digital machines and the internet.
- *IIoT*: IIoT stands for the industrial Internet of Things, the extension of the Internet of Things (IoT) in industrial sectors and applications related

to manufacturing. The IIoT embodies industrial applications, such as robotics, medical devices and autonomous automated processes.

- *Big data*: Big data is usually a large set of unstructured or structured data that can be organized, compiled, stored and analysed to study patterns, trends, opportunities, associations and debacles.
- *Artificial intelligence* (AI): Artificial intelligence is the science of simulation of human brain thinking patterns and intelligence processes using machines – particularly computer systems. Artificial intelligence is a built-in computer ability to do designated tasks and make decisions.
- *M2M*: This stands for machine-to-machine; this is the communication that happens between two disparate machines or systems via wireless or wired networks.
- *Digitization*: Digitization means creating a digital format of physical objects or attributes. It is a process of collecting and converting different types of information into a digital format to be stored in computers.
- *Smart factory*: A smart factory is one which uses Industry 4.0 technology, solutions and approaches. It comprises of connected devices, subsystems, internet technologies, machinery and production systems to continuously collect, process and share data.
- *Machine learning*: Machine learning (ML) is an AI that uses software for predicting accurate outcomes without being specifically programmed. Machine learning algorithms use the historical data as input and predict outcomes as required.
- *Cloud computing*: Cloud computing refers to the delivery of software and computing services, which encompasses storage, databases, servers, networking, software, analytics and intelligence via the internet.
- *Real-time data processing*: Real-time data processing refers to the capability of computer systems and machines to process data in a short period, providing near-instantaneous insights and output.
- *Ecosystem*: An ecosystem, in context of manufacturing, relates to the potential harmonization of the entire factory operation – inventory and planning, manufacturing execution, financials, HR, customer relationships, supply chain management and post-market vigilance.
- *Cyber-physical systems* (CPS): Cyber-physical systems is cyber manufacturing technology that offers "physical and engineered systems whose operations are monitored, controlled, coordinated, and integrated by a computing and communicating core".

In short, Industry 4.0 and its connected devices span over the entire product life cycle, manufacturing and supply chain – design, scheduling, quality, engineering and customers, producing much accurate, richer and timely analytics. The use of technology on the sub-processes within the manufacturing system will help in providing adaptive and predictive decisions.

17.8 THE SMART FACTORY – BUILT ON INDUSTRIAL IOT

The smart factory is the prime factor to Industry 4.0. It is a leap forward from a traditional automation process to a fully connected flexible and accessible system. It is an integral part to a broader digital supply chain network and has numerous aspects that manufacturers can leverage to adapt to the changing marketplace. The smart factory revolution will have an impact on velocity, scope and systems. It is an environment where machines and equipment will improve processes through automation and self-optimization.

The smart factory is characterized by automated workflows, synchronization of assets, improved tracking and scheduling. The optimized energy use inherent in the smart factory can increase yield, uptime and quality, as well as reduce costs and waste (*Source*: Deloitte insights).

Figure 17.3 depicts the transformation from traditional supply chain to smart digital factories.

Characterizations and descriptions derived from industry studies and reports describe smart factories by:

- Substantial improvement in manufacturing floor and their position in the supply chain.
- Fully connected and flexible systems, easy monitoring, waste reduction, redundancy check and improved speed of production are some of the advantages of automated manufacturing processes. It operates on the constant data flow from connected production and operations systems.
- The automation is beyond the traditional mechanization environment in a production facility, in terms of performing discrete tasks or processes.

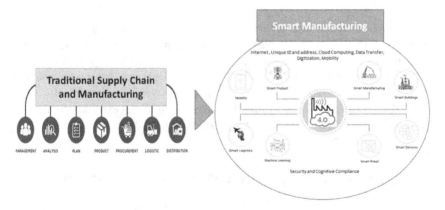

Figure 17.3 Transformation from traditional systems to smart manufacturing.

It is a platform built on an open and collaborative environment that allows selected partners to develop specific applications that extend offerings and reach to help deliver the promise of IIoT. The factory management is orchestrated via smart enterprise resource planning (SERP) systems and supported by human and virtual agents to develop products that are responsive in real time to demand, market conditions and value chain (e.g. logistics) feedback.

Smart factories have the underlying processes and materials well connected to generate the data necessary to make real-time decisions. To achieve this, the assets and machinery in a factory are fitted with smart sensors, which can continuously pull data sets from both new and legacy sources, ensuring data are constantly updated and analysed to reflect real-time conditions. Gathering and analysing data is a crucial requirement in the concept of the Factory 4.0, as it enables to free the right potential in the equipment, resources and people. The integration of data from operations and business systems, as well as from suppliers and customers, enables a holistic view of upstream and downstream supply chain processes, driving greater overall supply network efficiency.

Smart factory (manufacturing) allows factory managers to automatically collect and analyse data to make better-informed decisions and optimize production. In this environment, artificial intelligence (AI) will generate new insights from large amounts of data, and smart assets will operate autonomously. These applications will make complex networks more transparent for companies and enhance collaboration across corporate boundaries.

- The information from devices, sensors and machines are collected and communicated to the cloud by network and IoT connectivity at the factory-floor level.
- The collected data is sanitized, analysed and blended with contextual requirements and then reports are shared with authorized and lawful stakeholders.

The data collection and processing is designed to incorporate analytics to deliver pertinent and valuable business information. These applications and services include installation management / optimization, asset management and protection, tracking and condition based monitoring (CBM), predictive maintenance, augmented reality applications and calculation of overall equipment effectiveness (OEE). Value creation is impelled by data and analytics applications such as advanced optimization tools, machine learning algorithms and simulation software in all functions.

17.9 HOW IS THE INDUSTRY 4.0 MATURITY ASSESSED?

Maturity models are conduits of measurement and harmonizing a set of capabilities required to reach the desired maturity state. These models are premeditated as the comprehensible path represented by distinct maturity levels to reach coveted

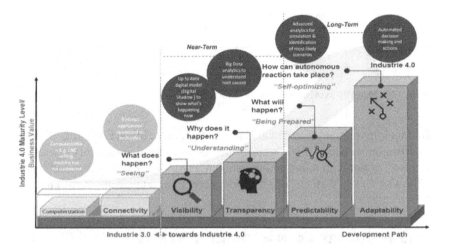

Figure 17.4 Industry 4.0 Maturity Model.

Source: Acatech Industrie 4.0 Maturity Model University of Texas, Dallas

objectives. These models have maturity assessments that will bring out the current organizational capabilities and maturity, adherence, limitations and gaps that need to be preset and enhanced before commencing the ontogeny of processes specifically defined by each maturity stage.

An evolutionary maturity assessment is used to measure the current maturity level of a certain aspect of an organization in a meaningful way, The assessments usually have a systematic literature review, interviews and probing and confirmations of the findings.

The following diagrams show Industry 4.0 maturity model.

17.9.1 Industry 4.0 maturity

The proposed model includes a total of six value-based development stages and four structural areas and 62 maturity items, which are grouped into nine dimensions. Before implementation and promotion of the maturity level, assessments are conducted using the following steps:

The content of the digital maturity model is constant, but the level to which an organization needs to be mature in each area is dependent on the organization's business strategy, business planning model and operating model.

The focus is on manufacturing enterprises and this assessment is aimed to be a guidance for organizations to develop their own strategy for Industry 4.0 (I4.0) implementation, which contains six value-based development stages.

Furthermore, Becker and Knackstedt (2009) explain that it is important for an organization to assess the maturity of a concept in order to continually improve

Figure 17.5 Maturity assessment – general methodology.

from their current position and also to highlight areas of development, allowing prioritization to occur. It is suggested that the organization uses this assessment results to not only to refine and mature but also share best practices and knowledge with its key partners, so that a long-term and comprehensive roadmap can be implemented.

17.10 INDUSTRY 4.0 IN MEDICAL SYSTEMS

In the medical field, Industry 4.0 displays extensive manifestation and an evolving marketscape. It has brought in a new technical capability of manufacturing customized implants, wearables and ingestibles, which can operate with utmost precision. The production of medical devices and implant is a big challenge for any manufacturer because every patient health information/data differs for specific health conditions. Medical device manufacturers are also experiencing increasing challenges in terms of price and margin pressure, speed to market, increased product complexity, manufacturing sophistication and more stringent regulatory and security conformity.

Industry 4.0 model is a transformation that embraces automation, security and generates new manufacturing opportunities in the medical world.

IoT and the Internet of Services (IoS) create a new ecosystem of connected facilities. It enables connectivity, integration and data exchange with the help of new technologies, software applications, embedded devices, censors, microcontroller, robots and other advanced network technologies. Industry 4.0 is evolving continuously, creating new opportunities and innovative sophisticated avenues for patient care. A visible part of the health care revolution has already manifested itself in medical instruments and procedures. For example, the cyber knife is used for delicate brain surgery and robotics perform prostatectomies.

It is plausible that Industry 4.0 toolkits will be employed more to health and social systems. With Industry 4.0 revolutionizing the face of medical technology, soon the integration of multiple devices into one will be the standard in the medical field. Similar to smart manufacturing, health care delivery has undergone a paradigm change and has got refined as "Health Care 4.0". Various patient-centric diagnoses and treatment options with customizations are continuously and exponentially being introduced. Extensive volumes of health data are generated and reported. Numerous equipment, sensors, actuators and devices have been installed and are well connected in hospitals, clinics, home, pharmacies and many other care environments for patient care.

Manufacturers of medical devices have unique requirements in terms of product details, throughput requirements, specificities, quality standards and adherence to regulatory intelligence guidelines. These particularities make medical device manufacturing a perfect candidate for adapting the Industry 4.0 paradigm. This transformation enables cost-effective, high-quality production of patient-specific or personalized medical care.

Another aspect of Industry 4.0 that is taking medical technology to the next level is additive manufacturing (AM). This involves the building of three-dimensional objects by layering materials, layer upon layer, until the desired outcome is reached. Another aspect of additive manufacturing is 3D printing, which is sometimes referred to as 3D bioprinting in the health care sector. Using 3D printing, the human tissues and organs are developed, which help scientists and researchers in clinical trials. Instead of the using animal subjects for trials, these simulated samples are used. Future implantable medical devices are a cost-effective alternative, promising a future of personalized patient care at an affordable cost.

Industry 4.0 encapsulates the benefits that come with developing interconnected cyber-physical systems without excluding legacy assets from the list. This brings in a variety of Industry 4.0 business models for medical device OEMs (original equipment manufacturers).

Traceability and control over the manufacturing process, as well as adherence and confirmation with regulations and standards, provide medical systems and devices manufacturers with the benefit of offering their customers high-quality products and outstanding service, ensuring customer safety and satisfaction.

Thuemmler and Bai state[6]: "The aim of Health 4.0 is to allow for progressive virtualization in order to enable the personalization of health and care next to real time for patients, professionals and formal and informal carers". The future of health care personalization is achieved by deploying CPS, cloud computing, the internet of everything, advanced data management and inferences from huge volume of data from worldwide patients. These historical inferences will not only aid in accurately identifying and diagnosing conditions but also determine safe and most effective treatment course of options and patient care for each individual.

Healthcare 4.0 is designed with a tendency to capture patient data and perform processing in applications, thus making health care management decisions predictive and better informed, while allowing for significant gains in efficiency and cost. The prospect of health care is now about preventing disease with AI, ML and data centred on patient behaviours. Doctors can ascertain conditions and can proactively and confidently recommend treatments to patients.

The Internet of Medical Things (IoMT) is also being implemented nowadays. It is an innovative way of amalgamating medical devices and health care applications, which include cloud-based platforms, applications, devices that will bring together volumes of discrete data from various sources and process such data.

Figure 17.6 Healthcare 4.0.

Reference: HC 4.0: www.sciencedirect.com/science/article/abs/pii/S2452414X19300135

Figure 17.7 depicts an "integrated service model" for care 4.0. This proposed Care 4.0 archetype views the person/patient holistically; screening their needs and aspirations for care services and connects the organizations that can support their needs. This Care 4.0 model heavily relies on a network of sensors, monitors and other health care technologies that constantly communicate to provide insights to managed services.

Care 4.0 / healthcare 4.0, which are nothing but industry 4.0, is now reaching new worlds with virtual reality, collaboration and remote support. For instance, the AR (augmented reality) technology allows the users to display information directly in the real-time environment captured through devices like smart phones, tablets, wearables or smart glasses. the aim is to create an integrated ecosystem that puts the patient at the centre and evolves towards more inclusive and collaborative therapeutic care models. Synchronized communication and sharing and processing of data and health information between health care facilities, doctors, care takers and patients are achieved.

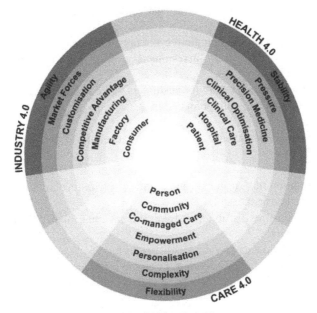

Chute & French, 2018
Visual Credit: Angela Tulloch

Figure 17.7 Integrated 4.0 Service Model.

Source: Chute & French 2019, Visual Credit: Angela Tulloch

17.11 REVOLUTIONS ARE DISRUPTIVE – ADVANTAGES AND DISADVANTAGES OF INDUSTRIAL REVOLUTION

The fourth industrial revolution is a term coined by Professor Klaus Schwab. He is the founder and executive chairman of the World Economic Forum, so he has some good credentials. He has described the fourth industrial revolution as a "current and developing environment in which disruptive technologies and trends such as the Internet of Things, virtual reality, robotics and artificial intelligence are changing the way people live and work".

Several advantages of the industrial revolution 4.0 – adoption of advanced technologies can be identified, namely:

- Increasing economic efficiency and strategic competitive advantage.
- Increasing labour productivity – increased knowledge sharing and collaborative working
- Flexibility and intelligence – agility

- Reducing manufacturing costs – lean with lesser redundancy
- Quick time to market – less machine and production line downtime, fewer quality issues
- Intelligence (AI) led predictive and preventive maintenance
- Increasing returns on investments – less resources, materials and product waste
- Improved quality of life for customers – like better patient care
- Lower barrier to entrepreneurship

There are also some caveats anticipated on account of adoption of advanced technologies.

- The diminution of human intervention – Non-recognition of machine-induced errors possible
- Core industries disruptions – Like cinemas being taken away by YouTube and Netflix
- Security and data privacy – cyber-security risk
- Skill development – continuous training is required for new technologies
- Different time frame for implementing Industry 4.0 with the integrated technologies.

The fourth industrial revolution warrants transforming manufacturing, health care and routine day-to-day lives by providing precise detection and personalized services. However, we have a risk that many IoT Systems are designed and implemented using diverse protocols and technologies that create complex configurations demanding greater security adherence. Also, there are significantly less reference and guidance for life cycle maintenance and management of AI technologies and IoT devices.

In the manufacturing industry, new technologies have been integrated in the production processes, resulting in pliable and modular production. Virtual and augmented reality in the health care sector – AR is used for surgery, medicine and rehabilitation. There is a vast demand for products and solutions that are able to improve current clinical practice.

Artificial intelligence in health care is a comprehensive, all-encompassing term used to describe the use of machine learning algorithms and software that helps in predictions and proactive actions. Connected ecosystems, clinical judgement or diagnosis, patient adherence and precision medicines are some of the AI and machine learning subjects of interest in Care 4.0 exemplar. Industry 4.0 has given birth to Care 4.0, which has impacted positively all categories of primary care, secondary care, tertiary care and quaternary care. Value-based system connected systems enables the health care Industry to improve in providing the quality of service (QoS) with well-informed decisions.

In conclusion, though Industry 4.0 arrives with bottlenecks, challenges and newer opportunities, these systems will impart positive changes to humanity and living by building a well-informed and an allied ecosystem. Huge data from devices breeds AI-driven solutions, which empower manufacturers to bring automation in manufacturing tools, machines, procedures, process and analytics. AI accelerates digital transformation and empowers the ability to run business smartly in this world of a connected ecosystem. Together, a new sophisticated revolutionary ecosystem is built and sustained.

Bibliography

Ahmed, S. T., Sandhya, M. and Sankar, S. (2019) A dynamic MooM dataset processing under TelMED protocol design for QoS improvisation of telemedicine environment. *Journal of Medical Systems*, 43(8), 1–12.

Ben-Israel, D. et al. (2020) The impact of machine learning on patient care: a systematic review. *Artificial Intelligence in Medicine*, 103, 101785. https://doi.org/10.1016/j.artmed.2019.101785

Bosch erweitert Portfolio für die Serializierung von Pharmaverpackungen, neue verpackung online, 26 January 2015.

Brettel, M., Friederichsen, N., Keller, M. and Rosenberg, M. (2014) How virtualization decentralization and network building change the manufacturing landscape: An industry 4.0 perspective. *International Journal of Science, Engineering and Technology*, 8, 37–44.

Buyya, R. and Dastjerdi, A.V. (2016) *Internet of Things. Principles and Paradigms*. Amsterdam: Morgan Haufmann / Elsevier.

Carroll, J.S. and Rudolph, J.W. (2006) Design of high-reliability organizations in health care. *Quality and Safety in Health Care*. 15 (Suppl 1), i4–9.

Chetna Nagpal, Prabhat Kumar Upadhyay, Annanya Chowdhury Bimal and Shubham Jain. (2019) *IIoT Based Smart Factory 4.0 over the Cloud*. IEEE. NSPEC Accession Number: 19379776.

Chute, C. and French, T. (2019). Introducing Care 4.0: An Integrated Care Paradigm Built on Industry 4.0. Capabilities. The Glasgow School of Art, Forres IV36 2SH, UK.

Culot, G., Nassimbeni, G., Orzes G., Sartor, M. (2020) Behind the definition of Industry 4.0: Analysis and open questions. *International Journal of Production Economics*, 226, 107617, https://doi.org/10.1016/j.ijpe.2020.107617

Evans, D. (2011) The Internet of Things: How the Next Evolution of the Internet Is Changing Everything. White Paper, Cisco, April.

Germanakos, P., Mourlas, C. and. Samaras, G. (2005) A mobile agent approach for ubiquitous and personalized ehealth information systems. In Proceedings of the Workshop on "Personalization for e-Health" of the 10th International Conference on User Modeling (UM'05). Edinburgh.

Gligor, D.M. and Holcomb, M.C. (2012) Understanding the role of logistics capabilities in achieving supply chain agility: a systematic literature review. *Supply Chain Management*, 17 (4), 438–453.

Grudzewski, F., Awdziej, M., Mazurek, G. and Piotrowska, K. (2018) Virtual reality in marketing communication – The impact on the message, technology and offer perception – Empirical study. *Economic and Business Review*, 4, 36–50.

Habib, M.K. and Chimsom, C. (2019) Industry 4.0: Sustainability and design principles. In Proceedings of the 2019 20th International Conference on Research and

Education in Mechatronics (REM), Wels, Austria, 23–24 May, pp. 1–8. doi:10.1109/REM.2019.8744120

Hood, L. and Flores, M. (2012) A personal view on systems medicine and the emergence of proactive {P4} medicine: predictive, preventive, personalized and participatory. *New Biotechnology*, 29(6), 613–624.

https://en.wikipedia.org/wiki/Fourth_Industrial_Revolution

https://public.wmo.int/en/bulletin/advanced-technologies-opportunities-and-challenges-developing-countries

www.computer.org/web/sensing-iot/contentg=53926943&type=article&urlTitle=what-are-the-components-of- iot

www.desouttertools.com/industry-4-0/news/503/industrial-revolution-from-industry-1-0-to-industry-4-0

www.merriam-webster.com/thesaurus/

www.mordorintelligence.com/industry-reports/smart-factory-market

www.ncbi.nlm.nih.gov/pmc/articles/PMC8367164/

Industrial revolution – from industry 1.0 to industry 4.0. www.desouttertools.com/industry-4-0/news/503/industrial-revolution-from-industry-1-0-to-industry-4-0

International Conference on Computer Science, Engineering and Education Applications. ICCSEEA 2021. Advances in Computer Science for Engineering and Education IV, pp. 401–410.

International Conference on Knowledge Management in Organizations – KMO 2017: Knowledge Management in Organizations, pp. 520–533. Cite as Design Science and ThinkLets as a Holistic Approach to Design IoT/IoS Systems

Internet of Things: Architectures, Protocols, and Applications, Volume 2017 |Article ID 9324035 | https://doi.org/10.1155/2017/9324035

ISC (2020) "Smart technologies" for society. *State and Economy*, 1397–1405 https://link.springer.com/chapter/10.1007/978-3-030-59126-7_153

Javaid, M. and Haleem, A. Industry 4.0 applications in medical field: A brief review, Current Medicine Research and Practice, https://doi.org/10.1016/j.cmrp.2019.04.001

Jingshan Li and Carayon, P. (2021) Health Care 4.0: A vision for smart and connected health care. *IISE Trans Healthc Syst Eng.* 11(3): 171–180. doi: 10.1080/24725579.2021.1884627

KPMG (2016) The factory of the future: Germany. Amstelveen, The Netherlands: KPMG AG. Available online: https://home.kpmg.com/xx/en/home/insights/2017/05/industry-4-0-its-all-about-the-people.html (accessed on 12 March 2018).

Kumar, S.S., Ahmed, S.T., Vigneshwaran, P., Sandeep, H. and Singh, H.M. (2021). Two phase cluster validation approach towards measuring cluster quality in unstructured and structured numerical datasets. *Journal of Ambient Intelligence and Humanized Computing*, 12(7), 7581–7594.

Lau, J.W.Y. (2020) Editor's perspectives – May 2020. *International Journal of Surgery*, 77, 218–219. doi: 10.1016/j.ijsu.2020.04.027.

Lobo, F.A. (2017). Industry 4.0: Redefining manufacturing of future medical devices. ECN. www.ecnmag.com/article/2017/10/industry-40-redefiningmanufacturing-future-medical-devices

Nakagawa, Elisa Yumi, Pablo Oliveira Antonino, Frank Schnicke, Rafael Capilla, Thomas Kuhn, Peter Liggesmeyer (2021) Industry 4.0 reference architectures: State of the art and future trends, Computers & Industrial Engineering, 156, 107241, https://doi.org/10.1016/j.cie.2021.107241.

A. (2021). *A Proactive Model to Predict Osteoporosis: An Artificial Immune System Approach*. Expert Systems.

Persico, V., Montieri, A. and. Pescape, A. (2016) On the network performance of Amazon s3 cloud storage service. In Cloud Networking (Cloudnet), 2016 5th IEEE International Conference on. IEEE.

Sathiyamoorthi, V., Ilavarasi, A. K., Murugeswari, K., Ahmed, S.T., Devi, B.A. and Kalipindi, M. (2021). A deep convolutional neural network based computer aided diagnosis system for the prediction of Alzheimer's disease in MRI images. *Measurement*, 171, 108838.

Schuh, G., Anderl, R., Gausemeier, J., Hompel, M. and Wahlster, W. (Eds.) (2018) *Industry 4.0 Maturity Index: Managing the Digital Transformation of Companies*. Munich, Germany: Herbert Utz Verlag.

Simchi-Levi, D. and Wu, M.X. (2017) Powering retailers' digitization through analytics and automation. *International Journal of Produciton Research*, 56, 809–816.

Thuemmler, C. and Bai C. (2017) *Health 4.0: How Virtualization and Big Data Are Revolutionizing Healthcare*. Cham, Switzerland: Springer.

Vochozka, M., Klieštik, T., Klieštiková, J. and Sion, G. (2018) Participating in a highly automated society: How artificial intelligence disrupts the job market. *Econ. Manag. Financ. Mark.*, 13, 57–62.

Yang, C.-T., Chen, L.-T., Chou, W.-L. and Wang, K.-C. (2010) Implementation of a medical image file accessing system on cloud computing. In Computational Science and Engineering (CSE), 2010 IEEE 13th International Conference on. IEEE.

Zhou, K., Liu, T. and Zhou, L. (2015) Industry 4.0: Towards future industrial opportunities and challenges. In Proceedings of the 2015 12th International Conference on Fuzzy Systems and Knowledge Discovery, FSKD, Zhangjiajie, China, 15–17 August, pp. 2147–2152.

Index

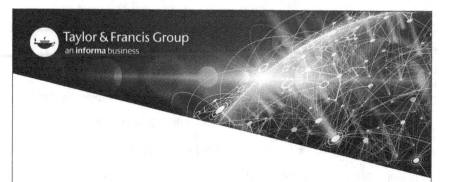

Taylor & Francis Group
an **informa** business

Taylor & Francis eBooks

www.taylorfrancis.com

A single destination for eBooks from Taylor & Francis
with increased functionality and an improved user
experience to meet the needs of our customers.

90,000+ eBooks of award-winning academic content in
Humanities, Social Science, Science, Technology, Engineering,
and Medical written by a global network of editors and authors.

TAYLOR & FRANCIS EBOOKS OFFERS:

A streamlined
experience for
our library
customers

A single point
of discovery
for all of our
eBook content

Improved
search and
discovery of
content at both
book and
chapter level

REQUEST A FREE TRIAL
support@taylorfrancis.com

Routledge
Taylor & Francis Group

CRC Press
Taylor & Francis Group